Silent Spill

Urban and Industrial Environments
Series editor: Robert Gottlieb, Henry R. Luce Professor of Urban and Environmental Policy, Occidental College

Maureen Smith, *The U.S. Paper Industry and Sustainable Production: An Argument for Restructuring*

Keith Pezzoli, *Human Settlements and Planning for Ecological Sustainability: The Case of Mexico City*

Sarah Hammond Creighton, *Greening the Ivory Tower: Improving the Environmental Track Record of Universities, Colleges, and Other Institutions*

Jan Mazurek, *Making Microchips: Policy, Globalization, and Economic Restructuring in the Semiconductor Industry*

William A. Shutkin, *The Land That Could Be: Environmentalism and Democracy in the Twenty-First Century*

Richard Hofrichter, ed., *Reclaiming the Environmental Debate: The Politics of Health in a Toxic Culture*

Robert Gottlieb, *Environmentalism Unbound: Exploring New Pathways for Change*

Kenneth Geiser, *Materials Matter: Toward a Sustainable Materials Policy*

Matthew Gandy, *Concrete and Clay: Reworking Nature in New York City*

Thomas D. Beamish, *Silent Spill: The Organization of an Industrial Crisis*

Silent Spill
The Organization of an Industrial Crisis

Thomas D. Beamish

The MIT Press
Cambridge, Massachusetts
London, England

© 2002 Massachusetts Institute of Technology

Set in Sabon by The MIT Press.

Library of Congress Cataloging-in-Publication Data

Beamish, Thomas D.
Silent spill : the organization of an industrial crisis / Thomas D. Beamish.
p. cm. — (Urban and industrial environments)
Includes bibliographical references and index.
ISBN 978-0-262-02512-6 (hc. : alk. paper) — ISBN 978-0-262-52320-2 (pb.)
1. Oil spills—Environmental aspects—California—Guadalupe Region—Public opinion. 2. Petroleum industry and trade—Environmental aspects—California—Guadalupe Region—Public opinion. 3. Pollution—California—Guadalupe Region—Public opinion. 4. Guadalupe-Nipomo Dunes National Wildlife Refuge (Calif.)—Environmental conditions—Public opinion. 5. Public opinion—California—Guadalupe Region. I. Title. II. Series.
TD196.P4 B43 2002
363.738'2'09794—dc21   2001044335

For my Mother and Father, who gave me the courage to try, the support to persevere, and the tools to finish.

Contents

Acknowledgments

I would first like to acknowledge generous funding from the University of California Toxics Research and Teaching Program and the US Minerals Management Service (contract 14-35-0001-30796).

Harvey Molotch's unflagging personal support for the research on which this book is based, and his intellectual mentorship, provided the bedrock for my efforts. Richard Flacks advised, taught, and encouraged me as a mentor and friend throughout my graduate career at the University of California at Santa Barbara in a caring fashion unusual in academia. John Mohr's candor, insight, and understanding of organizations sent me in directions that at first appeared daunting yet in the end provided the bases for two of my three empirical chapters. I thank John Foran for his willingness to read and comment on the final draft, and Nicole Woolsey Biggart for intellectual mentorship while I was a resident postdoctoral research associate at the Institute of Governmental Affairs at the University of California at Davis.

Without my colleagues at the University of California at Santa Barbara, the research and the book would have remained simple and incomplete. In two years, a number of investigators at the Ocean and Coastal Policy Center investigating the Central California Oil Industry evolved into a cohesive group that opened a space for critique and exchange concerning our individual research endeavors. Harvey Molotch was the leader. The intimates included Jacqueline Romo, Krista Paulsen, Leonard Nevarez, and, for a short period, the unfailing Andrew Lord. (This research was contracted through the US Department of the Interior's Mineral Management Services.) Others who contributed editorially and substantively include Todd Hechtman, Alec Tsongas, Brian Landau, Thomas Burr, Michael McGinnis, Lisa Torres, Paolo Gardinali, Britta Wheeler, Lynn Gesch, and Rani Busch.

The book also benefited from thoughtful responses, from citation leads, and from formal reviews of early papers and of the manuscript. Special thanks are extended to William Freudenburg, Diane Vaughan, Charles Perrow, John Jermier, and Robert Gottlieb. Moreover, by presenting papers and by participating in public forums, in projects, and research-sponsored quality review boards I subjected many colleagues to various findings and arguments. I would like to thank all those who listened, challenged, and expressed eagerness to see results, particularly Jim Lima, Loren Lutzenhiser, Ron Brieger, and Richard Walker. What is more, I owe a deep debt of gratitude to the people and organizations who allowed me the time to ask questions, let me take home enormous folders full of official and unofficial correspondence, and afforded me the use of back-room copiers.

Thom and Susan Beamish, my parents, not only encouraged my intellectual pursuits but also commented on drafts.

Jacqueline Romo, my partner in this adventure called life, provided insight, ideas, and constant support. To her I owe much more than this book.

Silent Spill

Introduction

There's a strange phenomenon that biologists refer to as "the boiled frog syndrome." Put a frog in a pot of water and increase the temperature of the water gradually from 20°C to 30°C to 40°C . . . to 90°C and the frog just sits there. But suddenly, at 100°C . . . , something happens: The water boils and the frog dies. . . . Like the simmering frog, we face a future without precedent, and our senses are not attuned to warnings of imminent danger. The threats we face as the crisis builds—global warming, acid rain, the ozone hole and increasing ultraviolet radiation, chemical toxins such as pesticides, dioxins, and polychlorinated biphenyls (PCBs) in our food and water—are undetected by the sensory system we have evolved.
—Gordon and Suzuki 1990

Underneath the Guadalupe Dunes—a windswept piece of wilderness[1] 170 miles north of Los Angeles and 250 miles south of San Francisco—sits the largest petroleum spill in US history. The spill emerged as a local issue in February 1990. Though not acknowledged, it was not unknown to oil workers at the field where it originated, to regulators that often visited the dunes, or to locals who frequented the beach. Until the mid 1980s, neither the oily sheen that often appeared on the beach, on the ocean, and the nearby Santa Maria River nor the strong petroleum odors that regularly emanated from the Unocal Corporation's oil-field operations raised much concern. Recognition, as in the frog parable, was slow to manifest. The result of leaks and spills that accumulated slowly and chronically over 38 years, the Guadalupe Dunes spill became troubling when local residents, government regulators, and a whistleblower who worked the field no longer viewed the periodic sight and smell of petroleum as normal.

The specific intent of this book is to relate how the change in perception took place, why it took nearly 40 years for the spill to become an agenda item (Crenson 1971), and why the response was controversial. The premise of the book is that social and institutional preoccupation with the "acute" and the "traumatic" has left us passive and unresponsive to festering problems. I begin with a general description of what locals have dubbed "the silent spill" (Bondy 1994).

I first heard of the Guadalupe spill on local television news in August 1995. (My home was 65 miles from the spill site.) The scene included a sandy beach, enormous earth-moving machinery, a hard-hatted Unocal official, and a reporter, microphone in hand, asking the official how things were proceeding. The interplay of the news coverage and Unocal's official response that caught my attention more than anything else. The representative asserted that Unocal had extracted 500,000 gallons of petroleum from a large excavated pit on the beach just in view of the camera. The newscaster ended the segment by saying (I paraphrase) "It's nice that Unocal is taking responsibility to get things under control." This offhand remark about responsibility set me to thinking about the long-term nature of the spill and about why it had not been stopped sooner, either by Unocal managers or by regulators.

A few months later, a colleague and I drove to the beach. My colleague, a geologist who was familiar with the area, had suggested that we visit the Guadalupe Dunes for their scenic beauty. We walked the beach and the dunes that border the oil field, alert for signs of the massive spill. The pit that Unocal had recently excavated had been filled in. The only hint of the project that remained was a small crew that was driving pilings into the sand to support a steel wall intended to stop hydrocarbon drift (movement of oil on top of groundwater) and the advancing Santa Maria River, which threatened to cut into an underground petroleum plume and send millions more gallons into the ocean.

Unocal security personnel followed along the beach, watching suspiciously as we took pictures. In fact, the spill was so difficult to perceive (only periodically does the beach smell of petroleum and the ocean have rainbow oil stains) that my impressions wavered. Was this really a calamitous event? The whole visit was imbued with the paradox of beauty and travesty.

Under my feet was the largest oil spill in California, and most likely the largest in US history. Table I.1 shows how large the Guadalupe spill is by comparing it with other notorious US spills. Yet the "total amount spilled" continue to be, as one local resident noted in an interview, a matter of "political science." There is still controversy over just how big this spill

Table I.1
US oil spills of more than 1 million gallons.

	Barrels	Gallons	Date
Guadalupe Dunes spill	476,190[a]	20,000,000	—[c]
(high and low estimates)	202,380[b]	8,500,000	
Exxon Valdez, Prince William Sound, Alaska (high and low estimates)	259,253[a]	10,900,000	March 24, 1989
	259,524[b]	10,100,000	
Burmah Agate, Galveston	254,761	10,699,962	November 1, 1979
Storage tank, Sewaren, New Jersey	210,000	8,820,000	November 4, 1969
Argo Merchant, Nantucket	183,000	7,686,000	December 15, 1976
Platform A well blowout, Santa Barbara Channel	100,000	4,200,000	January 28, 1969

a. high estimate
b. low estimate
c. This spill occurred over a period of 38 years.

really is. The smaller of the two estimates listed in the table (8.5 million gallons) comes from Unocal's consultants. State and local regulatory agencies do not endorse it (Arthur D. Little et al. 1996). The estimates quoted most often by government personnel put the spill at 20 million gallons or more, which would make it the largest petroleum spill ever recorded in the United States.

At first glance, it seems strange that so many individuals and organizations missed the spillage[2] for so many years; 'passivity' seems to be the word that best characterizes the personal and institutional mechanisms of identification and amelioration. It is also clear that the Guadalupe spill is very different from the image of petroleum spills that dominates media and policy prescriptions and the public mind: the iconographic spill of crude oil, complete with oiled birds and dying sea creatures.

The Guadalupe Dunes spill is only the largest *discovered* spill. Representing an inestimable number of similar cases, it exemplifies a genre of environmental catastrophe that portends ecological collapse.

Describing his impression of the spill in a 1996 interview, a resident of Orcutt, California, explained why he remained unsurprised by frequent diluent seeps: "When you grow up around it—the smell, the burning eyes while surfing, the slicks on the water—I didn't realize it could be a risk. It was normal to us." In a 1997 interview, a local fish and game warden—one of those initially responsible for the spill's investigation—responded this way to the question "Why did it take so long for the spill to be noticed?": "It is out of sight, it's out of mind. I can't see it from my back yard. It is down there in Guadalupe, I never go to Guadalupe. You know, I may have walked the beach one time, but I never saw anything. It smelled down there. What do you expect when there is an oil field? You know, you drive by an oil production site; you are bound to smell something. You are bound to."

In the days and weeks after my initial visit to the dunes, I wondered why the spill had gained so little notoriety. Beginning my research in earnest, I visited important players, attended meetings, took official tours of the site, and followed the accounts in the media.

What makes the Guadalupe spill so relevant is that it represents a genre—indeed a pandemic—of environmental crises (Glantz 1999). Collectively, problems of this sort—both environmental and non-environmental—exemplify what I term *crescive troubles*. According to the Oxford English Dictionary, 'crescive' literally means "in the growing stage" and comes from the Latin root 'crescere', meaning to "to grow." 'Crescive' is used in the applied sciences to denote phenomena that accumulate gradually, becoming well established over time. In cases of such incremental and cumulative phenomena (particularly contamination events), identifying the "cause" of injuries sustained is often difficult if not impossible because of their long duration and the high number of intervening factors.[3] Applied to a more inclusive set of social problems, the idea of crescive troubles also conveys the human tendency to avoid dealing with problems as they accumulate. We often overlook slow-onset, long-term problems until they manifest as acute traumas and/or accidents (Hewitt 1983; Turner 1978).

There are also important political dimensions to the conception of crescive troubles. Molotch (1970), in his analysis of an earlier and more infamous oil spill on the central coast of California (the 1969 Santa Barbara spill), relates a set of points that resonate with my discussion. In that article, Molotch examines how the big oil companies and the Nixon administration "mobilized bias" to diffuse local opposition, disorient dissenters, and limit the political ramifications of the Santa Barbara spill. Two of his ideas have special relevance: that of the *creeping event* and that of the *routinization of evil*. A creeping event is one "arraigned to occur at an inconspicuously gradual and piecemeal pace" that in so doing diffuses consequences that would otherwise "follow from the event if it were to be perceived all at once" (ibid., p. 139). Although Molotch is describing the manipulation of information for political purposes, his account of attention thresholds and of the consequences that the "dribbling out of an event" can have on popular mobilization resonates with both the "real" incident (i.e., the leaks themselves) and the "political" incident (the court case, the media coverage, etc.) that unfolded at the Guadalupe Dunes. Molotch's idea of the routinization of evil pertains to naturalization processes whereby an issue takes on the quality of an expected event and in so doing loses urgency. (What is one more oil leak if oil leakage is the norm?)

Our preoccupation with immediate cause and effect works against recognizing and remedying problems in many ways. It is mirrored in the way society addresses the origin of a problem and in the way powerful institutional actors seek to nullify resistance and diffuse responsibility. The courts and the news media, for instance, often disregard the underlying circumstances that led to many current industrial and environmental predicaments, focusing instead on individual operators who have erred and pinning the blame for accidents on their negligence (Perrow 1984; Vaughan 1996; Calhoun and Hiller 1988). Yet this ignores the systemic reasons why such problems emerge. In short, most if not all of our society's pressing social problems have long histories that predate their acknowledgment but are left to fester because they provide few of the signs that would predict response—for example, the drama associated with social disruption and immiseration.

Specific to pollution scenarios, in California 90 percent of marine oil pollution is attributable to unidentified, small, chronic petroleum releases that are neither investigated nor remedied. According to some experts, these smaller, less dramatic spills are "more severe than catastrophic [spills]" (Elliott 1999, p. 26). What is more, while legislation to stop dramatic tanker spills has halved the incidence of such spills off California, less dramatic spills on land continue unrestrained at 700 times the rate of tanker spills (Dinno 1999). Similarly, in 1980 the federal government officially listed 400,000 previously unacknowledged toxic waste sites across the United States; by 1988 the number had grown to more than 600,000. Of these, the

Environmental Protection Agency has designated 888 as highly hazardous and in need of immediate attention; 19,000 others are under review (Edelstein 1988; Hanson 1998; Brown and Mikkelsen 1990; Brown 1980). Recent estimates put the number of US sites with dangerously polluted soil and groundwater alone at more than 300,000 and the annual projected cleanup bill at $9 billion (Gibbs 1999).

Another example may provide some clarity, conceptually connecting instances that at first glance may appear disparate and unrelated. More familiar, but just as crescive and troubling, is the increase in ultraviolet radiation due to deterioration of the ozone layer. This has been "collective knowledge" for some time. Many of us have altered our behavior. More important, however, we have expanded what is normal to us by accommodating this looming threat. Applying sunscreen or avoiding direct sunlight has become routine. This is not, however, a solution; it is a coping strategy.[4] Would many people passively accept ozone depletion if cancer were to manifest in days rather than years?

The inability of our current remedial systems, policy prescriptions, and personal orientations to address a host of pressing long-term environmental threats is frightening. There are, however, numerous examples of disconnected events—seemingly unrelated individual crises recognized after the fact—that have received widespread public attention. Through national media coverage, images of ruptured and rusting barrels of hazardous waste bearing the skull and crossbones have become icons that fill many Americans with dread (Szasz 1994; Erikson 1990, 1994). But these are only the end results of ongoing trends that have been repeated across the country with less dramatic consequences. In view of the startling deterioration of the biosphere, much of which is due to slow and cumulative processes, more attention should be devoted to how such scenarios unfold. That is precisely what I intend to do in this book, in which I reconstruct how the parties involved in the Guadalupe Dunes case understood and responded to the chronic leaks.

Social scientists across the spectrum of interests agree that human action and interpretation can be made meaningful only by relating them to their social contexts. Like more conventional sociological topics, oil spills (Clarke 1990, 1999), toxic contamination (Mazur 1998; Brown and Mikkelsen 1990; Levine 1982; Brown 1980), and conflicts over industrial siting (Couch and Kroll-Smith 1994; Freudenburg and Gramling 1994; Edelstein

1993, 1988) are cases in which the objectives of industry, government, and the community structure the interpretation of the event, the range of solutions entertained, and ultimately the solutions chosen. In a similar vein, I focus on the Guadalupe spill's social causes and social ramifications and on the social responses to it.

My specific intent is to uncover how and why the Guadalupe spill went unrecognized and was not responded to even though it occurred under unexceptional circumstances. The industrial conditions were quite normal, and the regulatory oversight was typical. It would seem that there was nothing out of the ordinary, other than millions of gallons of spilled petroleum. This is, in part, why the spill is so instructive. It represents a perceptual lacuna—a blank spot in our organizational and personal attentions.

My approach stands in marked contrast to conventional environmental assessment, where analysis starts with the "accident" itself (i.e., post hoc) and moves forward in time and where the emphasis is on quantifying the direct impacts a hazard has had or is predicted to have on a localized environment.[5] The Guadalupe spill was not an accident and was a long time in the making. Tracing knowledge of the leaks as they worsened but were overlooked, ignored, and then covered up sheds light on "how contemporary disasters depend upon the way 'normal everyday life turns out to have become abnormal, in a way that affects us all'" (Hewitt 1983, p. 29). To this end, I trace the *career of knowledge* of the spill through its social contexts: the oil field (the origin of the spill), the regulatory institutions, and the local community. In each location, the search is for answers to the pattern of nonresponse. Why didn't local managers report the seepage, as the law requires? How did field personnel understand their role? How could pollution of such an enormous magnitude be left so long before receiving official recognition and action? Why did the surrounding community take so long to react?

It is important to underscore the exploratory and conceptual nature of my research. I use a particular case of contamination as an exemplar in an effort to better understand how human systems respond to critical and environmentally troubling scenarios. Slow-manifesting post-industrial accumulations of toxic substances present humanity with one of its greatest challenges. As Rachel Carson warned in *Silent Spring* (1962), they threaten the continued fecundity of the landscapes we inhabit and, by extension, our existence.

Ironically, the Guadalupe spill's crescive profile is revealed by the lack of a response to it. Because both organizations and individuals are preoccupied with spontaneously arising emergencies, they do not see problems of this sort until it is too late. Moreover, after such long gestation periods, and in view of the real constraints of feasibility and remedial impact, many of these contaminated sites present insoluble problems. Not only are they prohibitively expensive to "fix," but cleaning them up can be as destructive as leaving them as they are (Church and Nakamura 1993).

The organizations involved with the remediation of such environmental problems necessarily negotiate ecology, imposing human valuations on the environment in treating the impacts imposed through human (mis)use. Understanding this process of give and take (a sociological process, insofar as ecology is a non-hierarchical web of interconnected relationships) is crucial to developing a full view of societal intervention(s) (Shrader-Frechette and McCoy 1993). The official characterization processes (assessments of actual and potential damages),[6] while wearing the objective cloak of science, are applied by regulatory organizations and hired consultants whose agendas and responsibilities cannot be assumed to agree and are typically expressed in technical terms that limit inter-organizational (and inter-disciplinary) dialogue and interaction. Moreover, to reduce complexity and define causal relations, these analyses tend to "underdetermine" causal process in order to isolate aspects of the environment and determine cause and effect (Latour 1993). Hence, included in such reductionist formulations, but often left unarticulated, are the subjective underpinnings of environmental evaluations, which include assumptions concerning future use, idealized assessments of what is "natural," and determinations that differentially assess the importance of one medium relative to another (ocean vs. land, air vs. water, etc.).

Molotch (1970, p. 143) develops the notion of an *accident research methodology* in which the metaphorical accident is an occasion where the "breakdown in the customary order of things" lays bare just such previously hidden assumptions. Although in Molotch's example the disruption is quite sudden (an enormous spill of crude oil), the reasoning behind his use of this analytic strategy involves a great deal of crossover for a wide spectrum of social problems, including the Guadalupe spill. Molotch used the accident scenario to "learn about the lives of the powerful and the features of the social system which they deliberately and quasi-deliberately create" (ibid.). In the case of my research on the Guadalupe spill, the metaphorical accident—the 1990 recognition of significant petroleum contamination at the beach bordering the dunes—was a point from which to look both backward and forward in time and, in so doing, to gain entrance to the workings of individual and organizational rationality. It is because of the Guadalupe spill's position as a gray area between crisis and the customary order of things (Molotch 1970, p. 143) that the spill is so revealing a case.

Although sociological analysis of environmental phenomena is many times more widespread today than it once was, it continues to hold a peripheral position in mainstream environmental debates (MacNaghten and Urry 1995, p. 203). This is not to say that sociology or other social science work is unimportant. In fact, environmental concerns are a growing and increasingly important area within the social sciences. It is only to say that, in terms of "resources allocated, . . . the public visibility and acceptance of these works, and perhaps most of all . . . the attachment of this view to more powerful institutions of modern states" (Hewitt 1983, p. 4), the dominant paradigms concerning disaster, industrial crises, and environmentalism more generally lie in the physical sciences.

In a critique of the classical theories of sociology, Anthony Giddens (1990, p. 8) has gone so far as to assert that "ecological concerns do not brook large in the traditions of thought incorporated into sociology."[7] Historically, theorists of industrial societies, and before them theorists of agricultural societies, tacitly assumed the limitlessness of the environment and the limitlessness of human potential.[8] For instance, Marx (at least in his early writings) defined the human condition—particularly psychic health—in terms of man's ability to intentionally transform nature into the object of his desires (Marx 1974; McLellan 1977). Though Marx's insights into the contradictions inherent in capitalist systems of production and consumption are unrivaled, his attention to the industrial juggernaut's potential effects on the global ecological system was less than thorough or sustained. To Marx's credit, his writings, when painstakingly examined, do contain rudiments of what may be called environmental warnings (Dickens 1996; Foster 1999). For example, he developed a basic notion of soil nutrient depletion that he posited in large-scale industrial agricultural practice. Yet Marx and Engels articulated contradictory themes. On the one hand, Marx revealed the inherent contradictions that he felt would lead capitalism to destroy itself, of which agricultural soil depletion was just one manifestation. On the other hand, capitalism's inexorable global expansion meant that nothing in nature remained untouched. Nature, according to Marx, had become humanized. In view of current sentiments, this may seem to indicate that Marx and Engels were sincerely concerned with human domination of and penetration into everything "natural" (Merchant 1980). But that is not so. A strong component of Marx's writings was a theme that posits in the domination of "nature" the emancipation of human beings. Marx expressed the idea that a society that harnessed nature assured its members of freedom from the struggle to survive.

Durkheim touted an industrial age of interdependence and social fulfillment based on industrial expansion and division of labor. (See Durkheim 1984.) Moreover, Durkheim, with his early emphasis on explaining social phenomena exclusively by analyzing social facts by means of other social facts, actively eschewed the use of environmental factors to help explain human behavior. Until quite recently, sociology and social science more generally have, implicitly if not explicitly, advocated the idea that the human transformation of the environment was natural, unthreatening, even preferred. My point here is not to devalue the scholarship of Marx and Durkheim or to imply that rereading them and applying what one learns from doing so is fruitless; it is only to point out the intellectual "Balkanization of knowledge" and to emphasize the theoretical hole that is only recently beginning to be filled (Buttel 1987).

Mainstream sociology's historical neglect of environmental problems reveals a proclivity to sense only immediate and sudden threats to our well being (social or environmental). Especially in circumstances of slow and incremental change, threatening changes are normalized because actors (corporate and individual) accommodate themselves to gradually evolving signs of crises. This proclivity is not limited to environmental matters. For instance, Diane Vaughan's argument in *The Challenger Launch Decision* (1996) rests largely on the idea of normative drift—i.e., the idea that organizational actors, while working together, developed routines that blinded them to the consequences of their actions. Through their continual iteration, incremental expansion of normative boundaries took place, and unanticipated consequences resulted. This incremental expansion not only habituated social actors to what were in retrospect deviant events; over time it also increased their tolerance for greater levels of deviation. "Small changes . . . gradually become the norm, providing a basis for accepting additional deviance." (ibid., p. 409)

The response a potential threat receives depends largely on its social salience. However, contrary to intuition, salience is not always something obvious or easy to identify. For example, surreptitious forms of contamination such as radiation hold very little tangible and immediate effect; however, they can evoke a great deal of dread and awareness.[9] They provoke as much fear as earthquakes, floods, fires, hurricanes, or tornadoes (Erikson 1994). The defining feature of a threat, then, is its social salience, which captures the perceptual impact of a hazard's biophysical attributes and/or its social construction.

Thus, the salience of a crisis need not be derived only from extrinsic characteristics (e.g., a sudden onset, a dramatic and immediate impact). Salience also derives from less direct mediating social factors—factors in which a nexus of circumstances, both material and ideational, magnify perceived impacts—for instance, when a potential hazard affects many people (or, more important, when it affects politically endowed stakeholders) (Bullard 1990; Hofrichter 1993); when government responds swiftly and unequivocally (Cable and Walsh 1991); when daily routines are disrupted by an event (Flacks 1988); or, perhaps most significant, when the media define a hazard as newsworthy by providing for its widespread dissemination and problematization (Cable and Walsh 1991; Stallings 1990; Molotch and Lester 1975). These are all conditions that contribute to an event's salience. A conjunction of some or all of these factors can give an event notoriety even if it lacks obvious and immediate impact.

Low in immediate and tangible impact but high in public awareness, the events that surrounded the malfunction of a reactor at the Three Mile Island nuclear power plant in Pennsylvania are instructive as an example of political and media construction of social salience in a case where biophysical attributes were almost completely absent. On the morning of March 28, 1979, one of the two reactors at the Three Mile Island facility partially melted down, releasing radioactive steam into the surrounding countryside (Erikson 1994; Cable and Walsh 1991). Urging residents to remain calm, the governor suggested that pregnant women and preschool children evacuate an area within 5 miles of the plant. He also advised pregnant women and preschool children within a 10-mile radius of the plant to stay inside their homes. Unexpectedly, 150,000 men, women, and children—45 times the number of people advised to do so—fled the area. Although the Three

Mile Island incident lacked sufficient physical characteristics to impress local residents that something was wrong, it was quickly and unequivocally translated for them by regulators and other government officials. Moreover, extensive coverage in the national press lent it durability and drama that it otherwise may have lacked.[10]

At the other extreme is the 1989 *Exxon Valdez* tanker incident, in which an ocean-going oil tanker ran aground, disgorging as much as 10.8 million gallons of crude oil into Alaska's Prince William Sound. Though that accident occurred in a remote locale, it was sudden, obvious, and pictorially dramatic (Birkland 1998; Slater 1994; Clarke 1990). Its "media fit"—that is, its fulfilling the conventions of contemporary journalism (Gamson and Modigliani 1989; Wilkins 1987; Gans 1980)—also made it an extremely visible event. Virtually every major and minor news service in the nation carried copy and pictures as the story unfolded. And by disrupting the local commercial fishing industry, a crucial means of livelihood for the region, the event mobilized a group whose collective voice was hard for politicians to ignore.[11]

Industrial crises comparable to the *Exxon Valdez* and Three Mile Island debacles have gained widespread attention for similar reasons. The toxic contamination of Love Canal (Fowlkes and Miller 1982; Gibbs 1982; Levine 1982), the poisoning of the drinking water in Woburn (Harr 1995; Brown and Mikkelsen 1990), the abandonment of a dioxin-contaminated office building in Binghamton (Clarke 1989), and beaches turned black with crude oil near Santa Barbara (Molotch and Lester 1975; Easton 1972) are conspicuous examples of health-related crises that have garnered sustained attention from regulators and the public.

But lurking potential problems that currently lack extreme attributes, a convenient location, and an obvious beginning, and which do not lend themselves easily to media coverage, grow insidiously, getting little attention and rarely evoking an outcry. Pollution resulting from sea-bed disturbance, leakage of toxins from dumps, and deterioration of industrial infrastructure often present silent, slow, and creeping effects that accumulate incrementally over months, years, and decades, sometimes surfacing as catastrophes only after a long history of inattention and sometimes left entirely for future generations. Erikson (1991, p. 27) admonishes us to become aware of such phenomena and to act before it is too late:

Incidents of the kind [toxic contamination] that have concerned us here are really no more than locations of unusual density, moments of unusual publicity, involving perils that are spread out more evenly over all the surface of the earth. An acute disaster offers us a distilled, concentrated look at something more chronic and widespread. . . . Sooner or later, then, the discussion will have to turn to broader concerns—the fact of radioactive wastes, with half-lives measured in thousands of years, will soon be implanted in the very body of the earth; that modern industry sprays toxic matter of the most extraordinary malignancy into the atmosphere; that poisons which cannot be destroyed or even diluted by the technologies responsible for them have become a permanent part of the natural world.

The reality that surrounds crescive circumstances is characterized by polluters who are unlikely to report the pollution they cause, authorities who are unlikely to recognize that there is a problem to be remedied, uninterested media, and researchers who take interest only if (or when) an event holds dramatic consequence.[12] In short, all those who are in positions to address crescive circumstances are disinclined to do so. Forms of degradation that lack direct and immediate impact on humans, dramatic images of dying wildlife, or other archetypal images of disaster tend to be downplayed, overlooked, and even ignored.

The national print media certainly mirrored the propensity to ignore the Guadalupe spill (Hart 1995). Over the period 1990–1996, the national press devoted 504 stories to the *Exxon Valdez* accident and only nine to the Guadalupe spill.[13]

In a 1996 interview, a reporter for the *Santa Barbara News Press* offered his opinion as to why the Guadalupe spill had received little public attention until 1993. His view resonates with three of the four social factors articulated above (social disruption, stakeholders, and media fit):

We didn't see black oily crude in the water and waves turning a churning brown. We didn't see dead fish and dead birds washing up. We didn't see boats in the harbor with disgusting black grimy hulls. This is largely an invisible spill. It took place underground. . . . Because it was not so visual, especially before Unocal began excavation for cleanup, I think that it just didn't capture the public. . . . But after Unocal began excavations, driving sheet pilings into the beach, scooping out massive quantities of sand, setting up bacteria eating machines, burning the sand. It began to dawn on people the magnitude of this thing, but again it wasn't in their back yards, Guadalupe is fairly remote. . . . And it's not a well-to-do city [the City of Guadalupe]—comparatively, anyway, with the rest of our area. . . . So I don't think it really sparked the public interest as much as it could have or would have if it was . . . a surface spill.

Most discussions that have taken place on the subjects of the social causes and ramifications of chronic and widespread environmental despoliation have focused on the social construction of ordinary citizens' judgments. This is a consequential avenue for research to have taken, and it will also be pursued here. In addition to addressing community constructions, I will demonstrate why there are very good reasons to focus on the "risk perceptions" of "upstream" players, particularly in industry and in government.

Insofar as the Guadalupe spill goes back 38 years, one is tempted to write off much of it as a vestige of a "pre-environmental" era in which corporations, the government, and individuals were not conscious that dumping and spilling were detrimental,[14] and that similar events will no longer occur because we are now aware of the consequences. Two points of fact contradict such thinking. First, the Guadalupe spill was evident for at least 20 years in a time when popular consciousness concerning environmental issues was high and environmental laws were in place.[15] Second, regulators were concerned with the conditions at the Guadalupe field as early as 1982, and perhaps earlier (Ritea 1994; Paddock 1994a; Greene 1993a; Freisen 1993), but did not respond. We should expect similar incremental and cumulative environmental problems to continue to occur, even if environmentalism is rife. To be sure, negligence and criminal misconduct figure in the Guadalupe narrative, especially in the latter years. However, at least as important to the generation of destructive events is the interplay of selective perceptions, limited organizational attentions, personal stakes, and a propensity to accommodate socially and psychically low-intensity and nonextreme events.

Analytically conjoining the Guadalupe spill's attributes with the social contexts within which it occurred clarifies the reasons for 40 years of unattended leaks and spills. It becomes clear that simplistic explanations based entirely on operator error, corporate criminality, or governmental regulatory complicity all miss the mark. To address the context within which the meaning of the spillage evolved, I focused on social factors—for instance, where the pollutant originated, whether the pollution was the result of accident or negligence, where the pollutant ended up, who was affected by it, and whether, once discovered, it was remedied as quickly as it could have been.

Making sense of the Guadalupe spill entailed disentangling the discourses that constitute it as an issue. Analytically, this required tracking knowledge of the spill longitudinally through the social systems and the individuals that took part in the spill's creation at the oil field; the organizational reactions it received from federal, state, and local regulators; and the reception it received in the surrounding community once it became a public event.

The chapters of the book are ordered so as to parallel the movement of the spill, as an issue and as a real and growing problem, from one social and institutional setting to another. Thus, the book follows the career of knowledge about the spill, paying special attention to who knew of it first, what they did about it, and where the information went from there. The changing definitions of the leaks and the accumulating petroleum at the dunes—those that were in play before its discovery, those that were in play during the discovery, and those that were in play after the discovery—are traced through time. Methodologically, I explore these and the other social dimensions of the Guadalupe spill by means of an inductive approach to theory building that relies on multiple sources of data, not on any single source. I took this approach for two reasons. First, research in this area, and specifically on this topic, is so speculative that no metatheory exists against which to test observations. Second, my research interests made it necessary to cover the multiple settings, and hence the multiple sets of data, through which the discourse concerning the spill has proceeded.

Central to my research were field interviews with members of the local oil industry, government regulators, community members, and environmental activists. These interviews were tape recorded, transcribed, and systematically analyzed. In addition to the interviews, there were many spontaneous conversations—in hallways, in office waiting rooms, in the homes of those that were the intended interviewees—with individuals I had not originally contacted or planned to meet. Though not recorded, these conversations should not be seen as any less important than the others. I also pursued ethnographic context, recording scores of informal conversations concerning the spill. I accumulated and analyzed a substantial collection of archival materials, and I have followed media portrayals of the spill closely since 1989.

To understand the transformation in the meaning of the spill that occurred over 38 years, it is necessary to take notice of certain earlier developments inside and outside San Luis Obispo County. That is, the provenance of current conceptions that underlie impressions of the spill is simultaneously historically remote and politically proximate. The story of the spill thus includes the long-term presence of oil operations in the region, the demographic shifts that have taken place since World War II, and changes in sentiments toward industry and the environment.

Chapter 1 attends to the discovery of oil in San Luis Obispo County in 1864 and the social history. Chapter 2 addresses how environmental problems have typically been approached, how risk has been conceived, and some more recent trends in social sciences that specifically concentrate on the organizational constituents of industrial crisis. Chapter 3 looks at how the Guadalupe spill's organizational setting (both formal and informal) and the individual motivations of field operators made it possible to keep the spill secret for nearly 40 years. Chapter 4 tracks the redefining of the spill. Chapter 5 delves into the community's interpretations of the event and into how those who have been actively involved with the spill but who hold no official institutional positions have responded to the it and to its handling by regulators. In chapter 6, I apply notions of social organization, social stability, and social inertia to what I have reported in the analytic chapters in order to address the genre of environmental degradation or industrial crisis represented by the spill. Adding my observations to those of others, I develop a model of social accommodation that helps to explain how (in Diane Vaughan's words) "good people do dirty work" and why human systems seem inclined to wait "as the temperature rises" and "do nothing . . . dazed and complacent with the increase in heat."

1

Oil History, Oil Production, and the Guadalupe Oil Spill

Wildcatters have punched wells from San Simeon to Nipomo, from Paso Robles to Shell Beach. . . . Maps on file with the state Division of Oil and Gas show the county is riddled with abandoned dreams and would-be oil tycoons.
—Stover 1989a

The first prospects for oil in San Luis Obispo County followed closely on the heels of the 1859 discovery of oil at Titusville, Pennsylvania. In fact, the first well in San Luis Obispo County was drilled only 5 years later (Olsen 1986, 1991). Oil exploration was sporadic through 1900, and local history recounts only moderate success in developing it profitably (Gidney et al. 1917).

Eventually prospectors found small petroleum deposits. The Arroyo Grande field, developed by local land owners and by oil pioneers from outside San Luis Obispo County, heralded a boom in optimism about the prospects for oil (Nevarez et al. 1996). The discovery of this field brought Union Oil (founded in California in 1890 and now known as the Unocal Corporation) to the county, along with some 30 other oil companies that hoped to exploit Arroyo Grande.

Oil fields in San Luis Obispo, including Arroyo Grande, are generally located on isolated ranches and agricultural lands. This has tended to concentrate oil revenues in the hands of a few wealthy landowners. Additionally, unlike other regions where minor operators have at times proliferated, a few large operators have also monopolized the oil business in the county (Beamish et al. 1998). The fact that only a few local residents made money in the oil business meant that the positive impacts of petroleum exploration and production did not significantly permeate the local economy. The county's history with oil is captured in a local news article: "Wildcatters have punched wells from San Simeon to Nipomo, from Paso Robles to Shell Beach. Most have produced more sweat than oil. 'We're not like Kern County, with oil wells all over the place,' said county planner Steve Eabry. . . . But that hasn't stopped people from trying. Maps on file with the state Division of Oil and Gas show the county is riddled with abandoned dreams and would-be oil tycoons." (Stover 1989a)

While the dreams of a few drove continued speculation and exploration, the negative aspects of oil and oil-related production, even early in the twentieth century, were at times points of contention for local residents. As early as 1910, locals complained of oil on the shorelines of Avila and Pismo beaches. The county eventually hired a special detective to investigate these complaints, and he found that Union Oil and several Port of San Luis officials were responsible for the dumping of tar, refuse, and residual products of asphalt and bitumen in the bay.

As critical to local community impressions of oil extraction as its "lack of a local payoff" has been San Luis Obispo's proximity to oil-rich neighbors. Producers in these other counties have looked at San Luis Obispo's ocean access and deep natural harbors as points from which to transport their oil. Throughout the history of San Luis Obispo County, producers in Kern County and Northern Santa Barbara County have used pipeline systems and other means to move oil to docking and storage facilities at Morro Bay and Avila Beach. These developments began around 1900 with the upswing in regional oil production. Increased demand for oil both inside and outside the country led producers to look more closely at San Luis Obispo.[1]

Eventually, as demand for oil increased with the transition to oil-powered sea-going vessels in the first decade of the twentieth century (Williams 1997), investment in storage, docking, and loading facilities became practical. Union Oil was the first to do so when it purchased a defunct port and storage facility in 1907 near Pismo Beach, 10 miles west of San Luis Obispo City. Union added to an existing pipe and storage system that connected its operations in Santa Maria and the San Joaquin Valley. Although on the surface this seemed to be a healthy sign for the local economy, the effects of this sort of development proved negligible over the long term. Unlike the initial site-preparation and construction phases of oil extraction, pipelines employed few workers to keep them running.

During this early period, San Luis Obispo's primary connection with oil was its potential as a regional "way station" for its neighbors. More oil would pass through the county's pipelines and be stored in its tanks en route to other places than would be pulled from its reservoirs (Beamish et al. 1998). This had important ramifications for future impressions of oil production and thus for oil work as an economic alternative in the region. Because San Luis Obispo never produced as much oil (or oil of as high a quality) as Kern County, Santa Barbara, or Ventura County, it never developed the positive cultural and economic connections to oil that those areas did (Freudenburg and Gramling 1995; Paulsen et al. 1996).

Changing Demographics

World War II increased the demand for oil dramatically, and with it the hopes that the local petroleum industry would prosper. Because of the central coast's strategic position relative to the US Navy's Pacific fleet, production at the Santa Maria Basin field, which borders San Luis Obispo County, accelerated. This had a ripple effect on San Luis Obispo's facilities because Union Oil, the producer of that field, also had its primary transport facility at Avila Beach. However, the increase in activity, like past oil development, was fleeting. When Avila's storage and transfer facility was recognized as vulnerable to Japanese attack, operations were relocated further inland to better-protected storage and refining facilities in the San Francisco Bay Area (Welty and Taylor 1958). This spelled the end of another cycle in the "on again, off again" promise that petroleum-oriented development was to have in San Luis Obispo County.

However, the war's impact on San Luis Obispo was felt in other ways that would have a great deal to do with the present character of the county, the residents that lived there, and eventually the reactions they would have to the Guadalupe spill. The war brought a temporary surge in population after the military establishment chose the central coast as a primary location for the training and embarkation of Pacific Theater troops.[2] What had been an isolated enclave suddenly opened to the "outside" world. Many military personnel stationed in San Luis Obispo experienced the county's natural and social amenities and resolved to return (Lee et al. 1977).

In the 1960s and the 1970s, the trend away from San Luis Obispo's prewar agricultural base intensified with the region's rapid population growth. As the population came to be dominated by professionals, service-industry workers, and government employees, a concomitant change took place in the cultural orientation toward the natural environment's uses, its value, and the community's identification with it. This economic and ideological shift has been the most pronounced in the coastal areas, where ocean resources and a lush mountain environment prevail (Nevarez et al. 1996). Defense of the "environmental quality of life" has come to play a decisive role in county politics and in the stance local regulators have taken against Unocal in pursuing a cleanup of the Guadalupe spill.

The early period's relevance to present-day interpretations of the Guadalupe spill lies in how that period sensitized locals to particular issues, especially those that pertained to protecting the region's natural environment. The changes in how county residents conceived of their central coast home transformed the region's relationship with oil and industrial pursuits. On the one hand, San Luis Obispo's history as only a marginal oil-producing region meant that petroleum would exert only a limited socio-economic influence (Beamish et al. 1998). Instead, the county remained primarily an oil throughway and a temporary storage point. On the other hand, the county increasingly attracted people who wanted to escape urban and industrial sprawl. This conflicted with the development of the petroleum infrastructure. Petroleum development had little regional support. With time, any support it may have once had has disappeared.

The Setting

The Guadalupe Dunes are located in southwestern corner of San Luis Obispo County, 30 miles south of the city of San Luis Obispo (population 41,950) and 65 miles north of the city of Santa Barbara (population 86,000), bordering the Pacific Ocean. The closest population centers to the spill site are the cities of Guadalupe (population 6200; 3 miles inland from the site), Santa Maria (population 61,500; 10 miles due east), and a string of seaside towns to the north. Topographically, the field is bounded on the south by the Santa Maria River estuary and wetlands, on the north by a preserve managed by the Nature Conservancy, and on the east by agricultural lands.

Figure 1.1 provides a glimpse of the dunes' wilder side. Figures 1.2–1.5 illustrate the contrast between nature and industry that characterized the dunes until the cessation of petroleum extraction in 1990 and the subsequent decommissioning of the extraction infrastructure at the site.

Figure 1.1
Undeveloped Guadalupe-Nipomo Dunes.

Oil exploration and production began at the Guadalupe Dunes in 1947. The first commercial well, owned by Continental Oil Company, was finished in 1948. In 1953, Union Oil of California purchased the field and expanded operations. At its peak, in 1988, there were 215 producing wells yielding approximately 4500 barrels of crude oil a day (Arthur D. Little et al. 1997). The crude oil produced at the site, as is characteristic of oil from the central and south coastal areas, was extremely viscous, with a density resembling that of peanut butter. Union Oil used various methods to enhance the recovery of this oil; one of these was the injection of petroleum thinners into well bores to loosen the crude oil and to make it easier to lift the thick oil out and to pipe it.

"Diluent" (also referred to as "K-9 thinner") was the primary distillate used at the field to "dilute" crude oil (Stormont 1956). It was first used (and spilled) at the field in the 1950s.[3] Much like kerosene or diesel fuel, diluent is a combination of distillates. It is extremely difficult to identify definitively at the site. This is a point of great controversy for those defining the remedial options and for those prosecuting Unocal for the spill. Although diluent is refined and includes chemical additives (and thus is more "artificial"

Figure 1.2
"Grasshopper" pumping unit, Guadalupe Dunes.

than crude oil), it is still difficult to distinguish from other petroleum spilled
at the site. This was important because Unocal, as with any contamination
event, was legally liable only for spillage that could be identified as its own.
Moreover, because of the inexact makeup of different "batches" of diluent
used at the field over the 38 years of production and the natural oil seeps
found in and around the region, the difficulty of identifying "the problem"
(at least, according to the rules as they apply in a legal proceeding) is apparent. It is virtually impossible to ascertain precisely (and thus to prove in an
adversarial courtroom setting) how much was spilled, when the spillage
occurred, or who is entirely responsible. In fact, according to toxicologists
involved with the spill, that is scientifically impossible in view of the number of confounding variables.[4]

Diluent's current and future physical effects on the site and on the surrounding region are also contentious. Diluent contains carcinogenic chemical solvents, including benzene, toluene, xylene, and ethylbenzene (often
referred to as BTEXs), that make it potentially more toxic than crude oil
(McKee and Wolf 1963). However, its degree of toxicity and how long it
continues to be toxic when released into the environment remain contro-

versial. Estimates of how long the diluent at the site will take to break down biologically range up to 10,000 years if it is left alone.

According to a few optimistic appraisals (including Unocal's), the diluent poses no threat at all. In an interview, a Unocal supervisor corrected me when I referred to the diluent as a toxin. Instead, he admonished, it is a contaminant—something very different. A contaminant, I was told, is a foreign agent (not natural to the environment) that is not necessarily dangerous, whereas a toxin is necessarily dangerous. According to this scenario, a period for biological breakdown is moot, insofar as diluent presents no danger to the surrounding environment. Since diluent contains BTEXs, this interpretation holds little weight with local, state, and federal regulators or with the surrounding community.

The Guadalupe oil field encompasses approximately 3000 acres (6 square miles) of largely undeveloped oceanfront land, 60 percent of which is now believed to be contaminated with diluent (California Coastal Commission 1999). The contaminated area contains significant dune formations and continues to support an array of rare plant and animal species. The field is within the 18,000-acre Guadalupe-Nipomo Dunes complex, the largest uninterrupted coastal dunes complex in the western United States. The Department of the Interior has designated this biome a National Natural Landmark, describing it as follows (Arthur D. Little et al. 1997):

... the largest relatively undisturbed coastal dune tract in California. Five major plant communities are well represented and the flora exhibit the highest rate of endemism of any dune area in Western North America. Dune succession is exceptionally well displayed. No comparable area in the Pacific Coast possesses a similar series of freshwater lagoons and lakes so well preserved, with minimal cultural intrusions and harboring such great species diversity. The area serves habitat for both rare and endangered plant and animals besides being one of the most scenically attractive areas in Southern California.

Although the area is not densely populated, it is ecologically diverse. It is home to 12 federally recognized endangered and protected plant and animal species.[5] Fishing provides locals with an inexpensive source of food and a small group of commercial surf fisherman with a source of income.[6] However, little commercial fishing is now done, as the fish are said to have taken on a strong petroleum odor and taste (warden, California Department of Fish and Game, interviewed in 1997). The area is also used for recreational activities (surfing, hiking, bird watching, etc.).

Figure 1.3
Abandoned pipe, Guadalupe Dunes.

Underneath the dunes (where the diluent spill primarily resides) are extensive groundwater reserves. According to California's Regional Water Quality Board, these are future water resources for human consumption, as well as being vital to the area's continued ecological integrity. And, like other aspects of the spill, whether the diluent has affected or will affect the aquifer that lies under the water table is still the subject of debate. Unocal says it will not, environmentalists claim it will. Government agencies are unwilling to commit themselves to an opinion, owing to a lack of "proof."

Because the area sits at the coastal junction of San Luis Obispo County and Santa Barbara County, jurisdiction over the spill was initially confused. However, a majority of the oil field is in San Luis Obispo County, which gives that county jurisdiction over litigation and coordination of cleanup activities. Santa Barbara County has taken an advisory role. Although both counties have strong traditions of environmental activism, particularly regarding oil and ocean resources (Nevarez et al. 1996; Molotch and Freudenburg 1996; McGinnis 1991), the initial investigation and criminal charges were dismissed on December 21, 1993. The presiding judge granted a defense demurrer on the ground that prosecutors had filed criminal charges 2 days after a one-year statute of limitation had elapsed (Wilcox 1994b; Pfaff 1994). Civil charges were re-filed against Unocal in January of 1994 by the San Luis Obispo District Attorney. Two months later the case was settled out of court through a plea bargain that included a $1.5 million settlement, no admission of guilt, and the dropping of all 28 criminal charges against six employees. Ten days later, on March 25, the California State Attorney filed a civil suit against the company for failing to self-report its spills. The case was settled in July of 1998 for $43.8 million (Finucane 1998; Cone 1998), again out of court.

The Spill as Public Event

As a substantial amount of research on the media, the public agenda, and the social construction of risk has concluded, the press plays a crucial role in defining "those occurrences which . . . are attended to as important enough to become a part of the public experience" (Molotch and Lester 1975, p. 236).[7] Because of the Guadalupe spill's relative isolation from an urban (or suburban) center and its lack of an acute set of impacts, press coverage was critical to the word getting out and framing the incident for the wider public. But before the big beach dig of 1994 even the San Luis Obispo County press—known in the region for its unusually strong support of an environmentalist agenda (Nevarez et al. 1996)—was slow to pick up the issue. Initially, at least, the spill was just not dramatic enough to get on TV or make the paper.[8] The director of a local environmental organization, who had sought to get more public attention focused on the spill, remarked: "It is really tough to get people interested in it." He remembered being asked by a TV reporter after a meeting concerning the spill "Why should we cover the spill; why should the public care?" The region's largest daily, the *Telegram-Tribune*, eventually dedicated a number of stories to the spill (162 between 1990 and mid 1997), but only after drama was associated with the event. The only significant press coverage of the spill came in 1993–94 with the lone spectacle associated with it: the emergency excavation of contaminated sand and "free product" (3 years after it was officially recognized) in order to stop the migration of petroleum into the Santa Maria River and the Pacific Ocean. Eighty percent of all coverage came after 1993.

The California Department of Fish and Game estimated that 30,000 gallons of petroleum leached into the river and the ocean during the winter of

Figure 1.4
Petroleum storage tanks, Guadalupe Dunes.

1993–94 alone. In August of 1994, on the basis of such estimates, Unocal, under orders from the US Coast Guard and the California Department of Fish and Game, began to dig an enormous pit on the beach bordering one of the extraction wells to remove petroleum that threatened a sensitive habi-tat.[9] In February the *Telegram-Tribune* reported that some 672,000 gallons of diluent had been removed from that pit and more was being pumped out each day (Wilcox 1994c).

The Guadalupe spill was not Unocal's only spill in the region. In 1989 a mixture of crude oil, gas, and asphalt had been discovered 25 miles north of the Guadalupe Dunes, under the seaside town of Avila Beach. As at Guadalupe, Unocal refused to acknowledge the magnitude of that spill for years, claiming that it was small and isolated. However, once forced to assess the extent of the contamination (which originated at its Avila Beach oil storage and transfer facility), Unocal found that 400,000 gallons of the mix had contaminated five square blocks of Avila Beach. Eventually, Unocal purchased and excavated most of the town.

Official recognition that there might be problems at the Guadalupe field date back to 1982, when California Coastal Commission expressed concern about oil production occurring so close to state tide lands.[10] In 1986 and 1988, significant petroleum releases were reported by community members to the authorities, yet they received little official attention.[11] In these early instances of "daylighting,"[12] slicks formed on beach sands, in the Santa Maria River, and on the ocean. As late as 1990—after repeated reports of foul odors at the beach, a whistleblower's anonymous telephone call, and obvious signs of petroleum on the beach—the reactions of regulatory authorities were tentative. (See appendix B.)

Local Unocal managers kept the 1986 seepage "in house" by announcing at a meeting of oil-field employees that the petroleum had originated from "a boat [that] went by and emptied its bilge" (field worker, interviewed in 1997). Field workers were told by their superiors: "If anybody asks, tell them to talk to the superintendent." (US District Court, Northern District of California 1994, p. 3) In 1988 managers reported petroleum on the beach bordering one of the extraction wells, but Unocal averted extensive investigation by claiming that the spilled thinner was a vestige of previous oil production by other oil firms (Unocal Corporation 1994b). To back up that claim, consultants hired by Unocal confirmed for state inspectors that the distillate's "fingerprint" (i.e., its chemical composition) did not match what Unocal used in its current operations (California Regional Water Quality Control Board 1988).[13]

In 1990, local Unocal managers attempted once more, but with less success, to rebuff investigators, this time with claims that the product found on the beach, the river, and the ocean was an isolated occurrence (San Luis Obispo District Court 1993; Rice 1994; Friesen 1993). According to one of the first on-scene inspectors (lead staff member, California Regional Water Quality Control Board, interviewed in 1996): "So [an inspector] talked to them. . . . 'You got this oil field here. [The diluent] might be coming from the oil field?' . . . They said: 'We don't know where this is coming from. It is not the type of oil we use.'"

The origin of the spilled diluent (and some crude oil) seemed to coincide with an old oil sump[14] (known as "C-12") and another extraction well ("Leroy 5X") at the mouth of the Santa Maria River, which runs just south of the dunes and the oil field. This time, Unocal agreed to drill monitoring wells to assess the extent of the spillage. Those wells, monitored by the Regional Water Quality Control Board, began to slowly reveal the magnitude of the field's contamination. The same inspector quoted above recalled

Figure 1.5
Petroleum storage tanks, Guadalupe Dunes.

how the complexion of the investigation changed as regulatory authorities began to realize the extent of what had been until that time characterized as an "ordinary" spill:

I reviewed the quarterly reports as they came in. I got the first quarterly report and it said that they had been removing something on the order of 40 barrels of [diluent] per day. That is an astonishing amount of free product! Normally . . . you recover a huge amount in the initial stages of pumping, but if you look at it on a curve over time it just drops off exponentially. . . . In this case . . . every day for 90 days they pumped 40 barrels per day. It was a huge amount! So the next quarter report came out; they still maintained that amount. . . . And through the third quarter, again maintained that amount. At that point, I said: "There is something going on here that's not right. This is not an ordinary spill."

As the discrepancies accumulated, Unocal could no longer keep a lid on the spill. The beach, the wildlife, the groundwater, the surfers, and the marine environment all had advocates, however fragmented, who began to acknowledge something was wrong. As information of the spill slowly spread, agencies began to share information and to gradually respond.

In July of 1992 authorities received a second anonymous tip from a disabled field worker that the spilled petroleum was not accidental. He claimed that spills and leaks had been an ongoing problem at the field since the late

1970s (Paddock 1994a; Greene 1993b; Finucane 1992). The tip led officials to raid Unocal's regional field office in Orcutt, where they found maps describing the breadth of the site's contamination. Unocal's local managers had not yet admitted that these maps existed, much less shared them with authorities (Sneed 1998).

Then, with the heavy winter rains of 1993, the water table rose to such an extent that petroleum that had been partly hidden underground surged to the surface, coating beach sands and ocean waters. This time regulators could not overlook the sheen, the foamy vestiges, and the stench of petroleum. "Unocal finally began a concerted effort to identify the scope of the problem in 1993," reads an August 1995 San Luis Obispo Planning Department report—but only after officials returned to the field yet again. This time (winter 1993–94), however, the Coast Guard would invoke emergency powers of the newly enacted Oil Pollution Act of 1990 and call on Unocal to dig up the beach to try to block the diluent's migration into the ocean.

Under great pressure, Unocal guardedly acknowledged that discharges of diluent had been a recurring problem at the Guadalupe oil field as far back as 1985.[15] Yet company records seized by California Fish and Game wardens at their local offices agreed with the whistleblower's claims of significant spills and leaks further back (Paddock 1994a,b). However, because self-monitoring laws were not established until amendments to the Clean Water Act, Unocal was liable only for having failed to report its spillage from 1973 on.

Ironically, although it did not report the leaks as its own in the early 1980s, Unocal (as related above) had called and met with authorities about "petroleum of unknown origins on the beach" in 1988 and again in 1990. On both occasions, Unocal had reported petroleum on the beach but had denied responsibility. In the second instance, a field worker anonymously reported that the spills were Unocal's.

Whatever the eventual outcome of a future cleanup, it was painfully clear that the current forms of industrial self-monitoring—established by acts such as the federal Water Pollution Control Act of 1973, the federal Oil Pollution Act of 1990, the California Environmental Quality Act of 1970, the California Porter-Cologne Clean Water Act of 1969, and a myriad of California resource codes and regulations that relied on industrial self-reporting—had not been effective.[16] Self-reporting had not occurred, and

Figure 1.6
Efforts by Unocal to contain leaching diluent, Santa Maria River.

detection of the spillage came decades too late to avert tremendous accumulations of petroleum in the groundwater of the area and its subsequent and continued leakage into the bordering wetland, riverine, and marine environments.

In view of the history related above, significant questions remain concerning why the spill went on so long without self-reporting, why environmental enforcement did not occur, and why the eventual enforcement took the forms that it did. The overarching question is this: Why did neither Unocal nor the government agencies nor the surrounding community feel compelled to respond until calamitous circumstances were manifest? The answers turn out to be surprisingly mundane, and for that reason the threat they represent is all the more frightening.

2

Conceptual Footings

The quantification of nature, which led to its explication in terms of mathematical structures, separated reality from all inherent ends and, consequently, separated the true from the good, science from ethics. . . . Values may have a higher dignity (morally and spiritually), but they are not real and thus count less in the real business of life—the less so the higher they are elevated above reality.
—Marcuse 1964, p. 146

Framing the Environment: Asking and Answering Questions of Environmental Significance

Research and commentary on the environment have proliferated since about 1970. Measured by volume, the discourse on the environment is now virtually unrivaled among public issues. Nonetheless, the amount of attention it receives tells us little about how environmental crises are spoken of generally and substantively. These are questions of quality (*what* is said), not quantity (how many times it is said).

What form, then, has attention to environmental matters taken? How do we investigate, and how do we answer, questions concerning, socio-environmental interactions and consequences? Of particular importance for this discussion: What frames of reference and what assumptions underlie how we manage an extensive array of interlocking social and environmental problems?

Philosophically, the early epistemological roots of science have much to do with the continuing analytic bifurcation of society and nature.[1] Formative and illustrative of this are the early conceptualizations of Galileo Galilei (1564–1642), René Descartes (1596–1650), and Francis Bacon (1561–1626).

Late in the sixteenth century, Galileo set an important precedent with his pioneering mathematical descriptions of the motion of the celestial bodies. His claim that science should restrict itself to the study of "essential properties" of the material world (shapes, numbers, and motion) continues to pervade the sciences (Boorstin 1985; Randall 1976).

Building on Galileo's contentions, Descartes's formal seventeenth-century analytic split of mind and body separated the physical, obdurate outer world from an inner, soulful one. In dividing the world into *res cogitans* (mind; God) and *res extensa* (body; nature), Descartes ascribed to the latter mechanical and objective processes that he claimed were understandable only through mathematical description (Mumford 1934, 1964).

Bacon's contention that the goal of scientific inquiry was to control nature again reified and separated human systems from natural ones, imbuing human-nature relations with adversarial undertones (Merchant 1980; Marcuse 1964; Stanley 1968).

Forcefully addressed by Lewis Mumford in *Technics and Civilization*, the movement toward a mechanical view of the universe left the world populated by machines as metaphor and fact. The sensuousness of human subjectivity had no part in this cosmology. Mumford decried this "world-picture," believing that this simplification of reality was ultimately destructive: "By his consistent metaphysical principles and his factual method of research, the physical scientist denuded the world of natural and organic objects and turned his back on real experience: he substituted for the body and blood of reality a skeleton of effective abstractions which he could manipulate with appropriate wires and pulleys. . . . What was left was bare, depopulated world of matter and motion: a wasteland. . . . If science presented an ultimate reality, then the machine was . . . the true embodiment of everything that was excellent." (Mumford 1934, p. 51)

Although these philosopher-scientists by no means align directly with current scientific endeavor, they, as exemplars, point to a legacy that continues to divide the social and physical sciences and their subjects of study. Simply put, "science" has addressed the natural and social worlds as distinct and separate systems of logic (or, in Mumford's language, as two separate machines). This inclination has conceptually bounded and codified a distinction that remains largely intact. Scientific approaches that integrate the "social world" with the "physical environment" continue to be outside orthodoxy (Freudenburg, Frickel, and Gramling 1995; Humphrey and Buttel 1982).[2]

Flowing from the analytic split enunciated above, questions of environmental health have typically fallen to the physical sciences for assessments of human and resource damages. In cases of industrial crises (for instance, oil spills and other toxic releases), experts are typically called in to assess and remedy identified harms (Hewitt 1983). Classifying and then ascertaining the extent of degradation to an environment, once it has occurred, is only an initial step, technically referred to as *characterization*. The subsequent step, *remedy*, involves bringing to bear scientific expertise (and the accompanying technological apparatuses) to cure detrimental impacts. Admittedly, describing scientific processes in such a mechanical and simplistic way does little justice to the power such analyses (and associated technologies) have had in dramatically altering and enhancing how we survive environmental threats. Nevertheless, formal environmental analyses have concentrated on post-production and consumptive excesses—for instance, effluent discharge—eschewing analysis that critically evaluates the underlying systemic relations that bring about these environmental problems. At their root, current efforts to address destructive environmental trends expose deep-seated assumptions about how environmental troubles should be addressed. Again, the intention is not to belittle the machinations of scientific appraisal, only to broadly outline its predisposition, and in so doing identify its current limitations.

Even in the social sciences, where revisionist and critical approaches to social problems are relatively common, the traditional society/nature divide and decontextualized references to socio-environmental processes persists despite strong criticism (Lash et al. 1996; Hilgartner 1992). According to MacNaghten and Urry (1995, p. 204), "sociology has generally accepted a presumed division of labor which partly stemmed form the Durkheimian desire to carve out a separate realm or sphere of the social that could be investigated and explained autonomously." In short, this disciplinary boundary work of marking out a distinct "social realm" assured researchers a defensible sociological position (Shove et al. 1998).

When "orthodox" sociology has entered the environmental arena, it has tended to ignore the interlaced character of socio-environmental interactions and outcomes in favor of explanations couched in wholly social terms (Woodgate 2000; Dunlap 2000, Dunlap and Catton 1979). Historically at least, social scientific theory has been in the habit of reifying a society/nature split by steadfastly clinging to what Dunlap and Catton (1979) refer to as the "human exceptionalist paradigm."[3] Stated succinctly, the exceptional-ist paradigm assumed that the "exceptional features of *Homo sapiens*—language, technology, science and culture more generally—made industrialized societies exempt from the constraints of nature" (Dunlap 2000, p. 21). Although this is understandable in view of sociology's empha-sis, it also limits our understanding of the environmental problems we con-front, which obviously cross the socially constructed boundaries we have invented to cleave the world into comprehensible segments (sociology, psychology, biology, chemistry, and so forth). Citing Frederick Engels's resistance to social theory that would address ecological concerns, Dickens (1996, p. 42) is less forgiving of contemporary sociology and its under-theorized conceptions of nature:

Environmental sociologists, over 100 years later, are still trying to contain their analysis within sociology itself. Giddens and Beck are two of the most influential contemporary sociologists working on environmental questions. . . . Giddens has recently stated that nature has "ended." He argues that in modern society nature is no longer a phenomenon external to human social life. . . . Beck seems to be assuming that all knowledge is no more than a social construction. Beck seems quite confused. On the one hand he is, like Giddens, saying that all forms of knowl-edge are under constant interrogation and challenge. On the other hand, his notion of a risk society must assume that there actually are real mechanisms out there causing real and likely environmental disaster. Both Beck and Giddens (and Beck in particular) tacitly rely on a realist ontology in order to make their points while simultaneously denying that such a form of knowledge can exist.

Dickens, however, does not dismiss either Giddens or Beck—and he should not, inasmuch as both have contributed valuable insights to our collective understanding of the "modern condition," regardless of these conceptual shortcomings. Thankfully, in a growing number of exceptions, the socio-political attributes of environmental struggles have been analyzed in con-junction with the environmental conditions that undergird their popular support. One example is Brown and Mikkelsen's (1990) articulation of pop-ular epidemiology.[4]

In one of the few conjunctive areas where physical and social scientists have collaborated, efforts to ascertain risk have typically focused on for-mally ascertaining the risks posed by a range of emerging technologies (Slovic et al. 1979, 1983). Such risk research originated in a number of con-cerns, including the recognition that growth included both positive and neg-ative social and ecological impacts (Schnaiberg 1994, 1980), the need for standards of safety in the regulation of new technologies (Shrader-Frechette

1991), increased public concern about these new technologies and a deteriorating environment (Szasz 1994), and the insurance industry's need for
accurate data on which to base premiums (Heimer 1988, 1985). The
entrance of the federal government into this arena as an "environmental
enforcer" generated a rush to understand the implications of "pricey" regulatory interventions (Colella 1981; Tierney 1999). Initial efforts were primarily developed by industry-backed "production science" and represented
formal economic and engineering performance-versus-cost models that
attempted to capture the tradeoffs associated with legislative agendas and
technological advances.[5] What initially emerged from these efforts, in the
1970s and 1980s, were risk-assessment models that, under the aegis of feasibility and societal benefit, sought to formally rationalize environmental,
technological, and developmental risks (Mitchell 1990).

The tone taken by the early risk assessors (mainly engineers or, in
Schnaiberg's language, "production scientists"), however, was less than
reassuring to laymen. Although this line of inquiry, through presumably
objective means, had concluded that the risks posed by nuclear technologies,[6] waste-disposal techniques (for instance, incinerators), cancer-causing
chemicals, and so forth were minuscule, public resistance to these technologies has been tremendous (Eckstein 1997; Epstein 1991; Edelstein
1988). Nevertheless, the public haranguing many of these projects and technologies received did not (initially, at least) lead researchers to reformulate
their positions or their assessment methods.[7] Rather, their faith in the valid-
ity of their evaluations became the baseline from which they asked why the
public was so prone to irrationally misinterpret the benefits and the low
risks associated with "self-evident" societal gains (Cohen 1985; Wildavsky
1979; Lowrance 1976; Tversky and Kahneman 1974; Starr 1969).

The most notable models included those that emphasized psychological
heuristics (Tversky and Kahneman 1974; Fischoff 1990; Slovic et al.
1979), those that emphasized probabilistic arguments (Cohen 1985; Starr
1969), and those that focused on economic utility. According to the first,
the public miscalculates modern hazards because mental processes are
inadequate to the task of calculating what are minuscule risks (for instance,
those posed by nuclear technologies). As such, the beholders of such risks
tend to employ "computational short cuts" that overemphasize particularly memorable (e.g., dramatic) events, in effect ignoring events that are
less vivid, even if they are more common (Tversky and Kahneman 1974,

1981). The second model suggests that the public irrationally conceives the risks associated with well-engineered and "safe" technologies and unthreatening chemicals. Proponents of that model point to the public's use of inherently unsafe technologies without fear (Cohen 1985; Starr 1969). One familiar contrast used to highlight such "irrational perceptions" pits the public's acceptance of the automobile (probabilistically a very unsafe technology) against its general unwillingness to live next to waste incinerators and nuclear facilities even though the probability of being harmed by living near such facilities is—according to the experts— infinitesimal. The third model emphasizes the economically "rational, if understandably selfish, response to facilities and technologies that may constitute local undesirable land uses (LULUs) . . . whatever their objective risks" (Freudenburg 1993, p. 911).[8]

Yet recent research focused on experts and complex-technical systems has found that even technical professionals—the "rational" risk assessors— also fail to predict or diagnose the crisis potential of the systems they operate (Perrow 1984; Mazur 1973, 1975). For example, those who work in the nuclear industry often overestimate the safety of nuclear energy (Slovic et al. 1979; Mazur 1991; Freudenburg 1988). In this and similar cases, the experts tend to concentrate on minutiae, while the public focuses on the "bigger picture" (Clarke and Freudenburg 1993, p. 71). The point here is not to argue the merits or faults of nuclear energy but to acknowledge the variability of interpretation.[9]

Although greatly criticized since the 1970s by some members of the scientific community and the public, probabilistic and quantitative risk assessments and related cost-benefit analyses of socio-economic/environmental tradeoffs still hold sway in many formal decision-making contexts (the courts, for example). Addressing problems and defining solutions require "proof" that the identified trouble is adversely affecting an environment (people, fauna, flora, and so forth) or that a proposed solution is effective.[10] This is a primary reason why formal models of cost-benefit analysis and risk assessment have gained ascendancy (Schnaiberg 1980): They present a systematic and formulaic basis for making difficult decisions. Without such systematic proof-appearing exercises,[11] formal rational systems of law and governance cannot make choices in a hyper-litigious society.[12] Herein lies the crux of a significant problem: What constitutes evidence in these decision-making contexts? As my analysis of the Guadalupe spill will illustrate, proofs and solutions are not necessarily the outcomes of manifest and formal rational organizational or scientific processes.[13] Assessments and proposed solutions are just as likely to be by-products of latent organizational dynamics such as organizational culture, organization problem solving behavior, and legislative preferences (Clarke 1999, 1989; Vaughan 1996; Perrow 1984).

A common suggestion voiced in many of these risk scenarios by people in affected communities is that more public participation in decision making should be institutionalized (Brown and Mikkelsen 1990; Clarke 1989; Levine 1982; Edelstein 1988). Though this has been a rallying point, and though it has been emphasized by many activists in struggles against official pronouncements of the "public good," the issue remains ambiguous. For example, even though public participation in federal and state Environmental Impact Statements and other environmental assessment processes have been legislated and are paid a good deal of lip service, they rarely affect policy construction or implementation processes (Eckstein 1997; Schnaiberg 1994, 1980; Beck 1992a,b; Brown and Mikkelsen 1990). In fact, the trend may be closer to the reverse: that as environmental problems and our understanding of them have become more complex, our collective reliance on technicians, technical means of analysis, and highly complex solutions has grown ever greater (Beck 1996, 1994, 1992a,b; Lash et al. 1996, 1994; Giddens 1994, 1991, 1990). Dependence on such expertise involves a double bind. Though these technologies and the capability to use them would seem to hold the only answer to a range of frighteningly real environmental trends, they relegate decision making to a select few. On the technocrat's shoulders rest contested, value-laden, and some say "tragic" choices that might be addressed through more democratic means.[14]

At the heart of my research on the Guadalupe spill is an effort to better understand how human systems continue down environmentally destructive paths even as corporate compliance, environmental legislation, and public awareness seem to increase. Recently enacted environmental regulations have made polluting a potentially expensive affair. Such measures were, implicitly at least, assumed to hold the promise of compliance and environmental health. This has not, as of yet, been the case. As Bauman (1992) articulated in a monograph aptly titled *The Solution as Problem*, "solutions" are not always the end of it. Bauman questions whether, in the effort to save the planet, we should blindly and arrogantly trust our luck to human techno-scientific systems of intervention whose negative potentials have already proved as dangerous as they are helpful. He calls on researchers to take their investigations inside the "machine" and to focus their attentions on such systems in the hope of revealing how solutions can lead to frighteningly real problems. In Bauman's words (ibid., p. 25): "The problem is not only that we are facing challenges on an un-dreamt-of scale, but, more profoundly, that all attempts at solution bear in themselves the seed of new and more difficult problems. . . . The most fearsome of disasters are those traceable to the past and present pursuits of rational solutions. Catastrophes most horrid are born—or are likely to be born—out of the war against catastrophes. . . . Dangers grow with our powers, and the one power we miss most is that which divines their arrival and sizes up their volume."

On reflection, then, the priority assigned to physical assessments (i.e., biological integrity, quality of surface water and groundwater, geomorphic composition, and so forth) and orthodox social science's penchant for staying on society's side of environmental issues have limited the directions of research, of discovery, and of solution-generating processes. Moreover, the increasingly complex and technical nature of the problems that societies confront and of the proposed remedies makes decision making difficult and contentious.

What is more, an excessive focus on aggregate outcomes rather than on process (i.e., how systems systematically reproduce environmental destruction) has also encouraged a form of myopia. Assessments of biodiversity, appraisals of ozone depletion and of the accumulation of greenhouse gases, economic projections of industrial growth in juxtaposition with estimates of the cost of environmental protection, demographic assessments of population growth and concomitant projections of food scarcity, and assessments of the risks associated with a range of scenarios and technological innovations are churned out unceasingly.[15] There is no doubt that this research is necessary. Nonetheless, the manner in which degenerative environmental trends are being addressed has left their a priori and systemic basis largely unattended, and this creates a nagging sense that everyone, but no one in particular, is creating environmental problems.

What needs more empirical attention is the question of how environmental problems are generated "on the ground"—for example, how actors enact, at the interpretive interpersonal and organizational levels, environmentally destructive trends. To answer questions of this type, we must look at the interlocking social and environmental relations that constitute environmental troubles as they are created.

Revisionist Research: Beyond Realism and Naive Social Constructionism

Rarely has there been research that has integrated how environmental crises manifest themselves, how they are made sensible and rationalized by society, and how they are acted on as an interactive process. Contemporary research in the social sciences, however, has begun to move beyond the naive positivism prevalent in the assessments of "human" and "natural" worlds, instead theorizing the mutually constitutive relationship between the two. (See, e.g., Freudenburg, Frickel, and Gramling 1995.) Couch and Kroll-Smith (1990, 1991, 1994) emphasize the nexus of ecological and symbolic factors in explaining how fundamentally disturbing ecological crises are to communities that endure them. Especially in cases where human agency is at the heart of a calamity,[16] rebuilding a sense of security and community is especially problematic for those affected. Residents grapple with questions of human intentionality and of blame, as well as with a nagging distrust of a now "tainted" environment.[17]

Unlike acute natural disasters, which have pronounced features and obvious effects that come and go seemingly by chance, industrial crises often occur without a clear set of warning signs or impacts and without a clear point of departure. Communities that must contend with human-induced crises—whether they are attributed to operator error or negligence, to failure-prone systems, or to some conjuncture of these—experience them as by-products of human intentionality. Thus, beyond the dread associated with not knowing whether one's health has been compromised, community outrage is directed at those who have unfairly instigated their duress. That is to say, both the physical and the social environment become tainted by traumas sponsored by human deeds. Addressing their interrelation is critical in accounting for impacts and understanding response.

In another example of research that attempts to address the complex and contextual nature of crises and interpretation, Mazur (1998) focuses on the hermeneutics of sense-making in the toxic event that unfolded at Love Canal (a chemical landfill that had become a residential neighborhood in the city of Niagara Falls, New York). Referring to conflicting accounts of the event metaphorically as the Law of Rashomon,[18] Mazur reconstructs the story of Love Canal from different interest perspectives. "At a minimum," he writes (ibid., p. 6), "the producers versus the victims of toxic chemicals [will have] conflicting accounts of what happened." Mazur's investigation, along with a growing body of revisionist risk research, makes the case that divergent interests, particularly personal values, are fundamental to variations in the judgments that a range of risks receive (Mazur 1998, 1975, 1973). This is not to say, based on the variability in interpretation, that one or another interpretation is irrational.[19] Again, this is the position more orthodox methods of risk analysis initially took when a large segment of the public rejected technologies and environmental standards that such methods had calculated as probabilistically risk free.[20] Such methods also discount community experience, a context out of which public trust in institutional forms emerges. Theorists such as Ulrich Beck (1996, 1992), Anthony Giddens (1991, 1990), and William Freudenburg (1988, 1992, 1993) have converged on the view that public impressions of institutional trustworthiness play a critical role in these interpretative processes.

Erikson (1994) and Edelstein (1993), discussing community response to hazards, argue that trauma often emerges from intense feelings of victimization, vulnerability, and stigma that erode confidence in societal institutions. In a similar vein, Freudenburg (1993, p. 927) found that institutional legitimacy and risk perceptions pivot on how institutions and "experts systems" carry out their duties "with [the] full degree of competence and responsibility that their fellow citizens" expect.[21] Freudenburg used the word 'recreancy' to characterize failure by an institution that holds a position of public trust to follow through on its duty (ibid., p. 916). Such failure generates public skepticism and undermines institutional authority (Wynne 1992, pp. 281–282).

Research that attends to the complexity inherent in risk scenarios is very important, for at one extreme the primary goal of the "realist school" of risk assessment (Starr 1969; Cohen 1985; Wildavsky 1979; Lowrance 1976) has been the (over)simplification of these circumstances through the search for objective points from which to measure how threatening a phenomenon is or will be. At the other end of such a hypothetical continuum, a radical social-constructionist perspective (i.e., social constructionism in toto), while generally sympathetic to community struggles, would claim that risk interpretation is wholly a subjective construction on the part of both risk takers (e.g., government and industry) and those resisting their claims (e.g., impacted communities).[22]

It is conceptually most productive to think of social-constructivist arguments as being of "strong" and "weak" varieties (Dittmar 1992). The strong version, as noted above, tends to deny the importance of the environment as an object external to human symbols and discourse. (For examples, see Tester 1991 and Haraway 1991.) In this, its most far-reaching form, "objective reality" is denied a place at the table. Even if objective reality exists "out there," this phenomenological outlook contends that we cannot experience it directly. Thus, "reality" is fruitfully dealt with as an ontological phenomenon rooted squarely and solely in the mind. The "weak version," although it recognizes that all knowledge is the outcome of social construction, does not go so far as to insist that the environment does not "exist outside of thinking and language" (Freudenburg 2000).[23] Contrary to strong social-constructionist notions, in "refashioning nature" society does not also change the "laws of nature" (Dickens 1996, p. 129). The point here is that the power of discourse over human understanding does not make us immune from dangerous environmental circumstances, regardless of the meanings imputed to them. It is, however, undeniably important to address how society represents through discourse and rhetoric its "rearrangement" of natural systems. Again, that understanding must also include how the physical environment structures, feeds back into, and refashions discourse and hence choice sets and decisions. Understanding this interplay is of the utmost importance if we wish to intelligently intervene and address a host of environmental problems.

The two social-constructivist positions, however, do coalesce on the point that it is misguided to give special privilege to technical-expert appraisals of risks, insofar as those appraisals are based on desires specific to the intentions of the "appraiser." Thus, decisions of risk should be made democratically so that the subjectivity of laypersons as well as that of experts is a part of decision-making processes.

As should be clear, what is needed is an integrated approach that theorizes symbolic and physical environmental qualities in simultaneity. By addressing how people "on the ground" (i.e., in interaction with an obdurate external reality) interpret, experience, and represent the threat of environmental disruption, we can get much closer to such an understanding.

Disruption and Disaster: Responding to Environmental Crises

As can be gathered from the foregoing, the risk polemic dovetails with research traditions that overlap but are focused on broader topics. One research tradition that has great relevance to the Guadalupe spill is *disaster research*. The examination of disaster, in the broadest of terms, currently falls loosely into two research regimens, based on the kinds of phenomena that are the focus of attention and the origin(s) of the threats described. These are the study of natural disasters with an emphasis on post-impact analysis and the emerging literature on human-induced crises. The emerging literature focuses on pre-impact or causal analysis and also on post-impact psychosocial affect. While the interests of this book emerges from the latter much of what currently appears as "disaster canon" and that is mirrored in formal institutional protocols flows out of the earlier post-impact emphasis of disaster assessment. For this reason, a review of this research tradition is fruitful.

The Cold War and Disaster Research
What I will refer to as a *post-impact model* has extensively documented how communities, organizations, and other social groupings respond in the aftermath of acute natural and human-induced disasters (Fischer 1998). The study of disasters gained a great deal of support during the Cold War as the threat of nuclear holocaust loomed. Disaster research provided a means of exploring how communities react to large-scale disruptions and thus of making inferences on the probable consequences of a nuclear attack (Turner 1978; Erikson 1976). The studies done since 1950, especially those emanating from the University of Ohio Disaster Research Center in the 1970s, specialized in post-impact analysis. They looked at what constituted effective communication strategies and at how myth and rumor functioned in these situations, and they recorded descriptions of all the forms that collective reactions and response took.[24] In view of the stated agenda, it is little surprise that this research was overwhelmingly post hoc, with "accidental" occurrences as the analytic focus and assumption (Hewitt 1983). The phenomena chosen for study had to be sudden, unexpected, and dramatic. Natural disasters such as earthquakes, tornadoes, floods, and fires lent themselves nicely to this research regimen's objectives.

A great deal of information has been gained about how to warn populations of imminent threat, about the social effects of "immediate-impact disasters," about how individuals react when caught in extreme situations, and about how much time it takes for a community to restore itself after a natural disaster. For instance, disaster research has formalized the stages, or "life cycle," of disaster scenarios. Each stage is discrete from those that proceed or follow in both behavioral and organizational terms. The *pre-impact* period entails some preparation, if time permits. The *impact* period is largely self-explanatory, but duration is important. As traditionally conceived, this is the shortest and most dangerous stage—the stage when "the hurricane blows through town." *Immediate post-impact* entails survivors' confronting their "new reality" after the calamitous event. The *recovery period* necessitates debris clearance and initial social reorganization. The *reconstruction* period is focused on the physical rebuilding of human infrastructures. (Fisher 1998, pp. 7–8) Though it is helpful for understanding extreme environmental conditions, this primarily post-event model does little to clarify threats and impending crises that are slow in onset and ambiguous in manifestation and that provide no end point on which to rebuild (both infrastructurally and psychosocially). Because these "new species of trouble" (Erikson 1994) often present ambiguous and insoluble profiles, they are generally consigned to long-term monitoring and management at best; at worst, they remain unidentified, unresolved, and in many instances actively denied by those who are responsible.

By counterpoising the two research endeavors at the conceptual level, we can get a clearer picture of the differences between their focuses. Comparing Charles Fritz's operational definition of *natural disaster* and Stephen Couch and Stephen Kroll-Smith's articulation of *human-induced crises* is particularly instructive for this purpose. Fritz (1961, p. 655) defines a disaster as "an event, concentrated in time and space, in which society, or a relatively self-sufficient subdivision of society, undergoes severe danger and incurs such losses to its members and physical appurtenances that social structure is disrupted and the fulfillment of all or some of the essential functions of society is prevented." According to Couch and Kroll-Smith (1985, p. 566), a chronic technologically induced hazard is "a slowly developing, extended, humanly produced deterioration in human-ecosystem relations, in which an entire community or sectors therein perceive and/or incur danger to health and safety and the disruption of ongoing patterns of social and cultural relations." Fritz's definition, though it captures the immediate impact scenario, leaves much unexplained when it is applied to events that have ambiguous parameters in both time and space (for instance, diffuse and indeterminate chemical contamination, which are often "non-extreme" in their initial manifestations—see also Erikson 1994). Research on post-impact analysis, until recently,[25] received little criticism for its short-sighted view of what constituted a "disaster" because the causes of nuclear attack were seen as exogenous (that is, out of the province of such research). As Turner (1978, p. 37) notes in his research on human-induced hazards, "the preoccupation of the military reinforced the more general perception of disasters as essentially unforeseeable, and encouraged the view that the problem of disasters was located only after the moment of impact."[26]

The post-impact model laid the groundwork for the subsequent study of less obvious phenomena. That model perseveres in manuals, protocols, and legislation that guide remedial action on the part of trustee authorities during local, state, and federal "states of emergency." Such ruinous trends as industrial decay, effluent discharge, toxic accumulations, and a range of other less immediately consequential and yet festering problems that lack a convenient beginning or end are passed over until they exhibit acute impacts of the sort that would trigger a response. That is, the ongoing reliance on the post-impact model at both the conceptual and the applied level loads the regulatory system against recognition and early prevention, relying instead on a "react and rebuild" mentality.

For a number of reasons, the post-impact model is increasingly suspect as we confront environmental hazards that cannot be addressed or solved in a post hoc manner. First, studying only the upheaval that follows an industrial crisis deprives us of one of the most important insights such an analysis can lend: information on what caused the problem, which might make recurrence avoidable. Second, because the nature of many post-industrial forms of spoliation (e.g., chemical and radioactive) involve an inability to fully restore an environment to its original form, a post-event analysis loses meaning. There is often no definitive beginning and just as often no real end. This is because many man-made wastes thoroughly penetrate an environment and are not "extractable" without dramatically altering the landscape in equally detrimental ways. DDT is a notorious example (Carson 1962), but cousin chemicals such as dioxin are of the same order.

Similarly, radioactive materials and chemicals such as trichloroethylene (TCE), polychlorinated biphenyls (PCBs), toluene, and benzene[27] can be extremely hazardous, and they are virtually impossible to remove once dispersed in an environment. Third, once toxins are removed, where can they be put: in storage containers? in landfills? where else? In view of the added expense, the threat of exposure, and the controversy that often accompanies these "ticking time bombs," removing them is problematic environmentally, fiscally, and emotionally for communities that are the intended destinations (Edelstein 1988; Hofrichter 1993). The threat they present never really ends, even artificially.

Anthropocentric Cause: Hazards, Hazards Interpretation, and Hazards Response

A research agenda that has emerged since the 1980s is broadly attentive to the human and organizational precursors of industrial crises, revisionist in tone, and still somewhat speculative in its conclusions. Nonetheless, this research is more pertinent to the Guadalupe spill (and to similar events) than traditional disaster research, since the pollution legacies of industrial and post-industrial landscapes are caused and experienced in qualitatively different ways than natural and acute disasters.

An important departure for this newer research on industrial crises has been its attempt to understand the truly disturbing quality a diverse range of hazards has on the human psyche and on the community. Why do some hazards capture the imagination while others do not? The social salience of surreptitious forms of contamination (e.g., radiation and toxic contamination), while having very little tangible and immediate affect, have great psychological impact on individuals who deem themselves affected (Erikson 1994). Again, perceptions of threat are socio-environmentally contingent on a range of constituent elements, including familiarity or unfamiliarity with a phenomena (Slovic et al. 1979), valuations of catastrophic potential (Perrow 1984; Tversky and Kahneman 1974), evaluations of personal efficacy and/or control over the object of anxiety (Freudenburg 1993, 1988), and value interests relative to a proposed technology or its perceived effect(s) (Mazur 1998).

Although research in this area is diverse, the revisionist school de-emphasizes post hoc examination of disasters' impacts and instead focuses on upstream causes and their downstream implications. For example, Kroll-Smith and

Couch (1991) and Erikson (1994) focus on human-induced trauma that is of the crescive and insidious type. Another intellectual stream, generically referred to as "normal accident theory," is especially relevant to the events as they unfolded at the Guadalupe Dunes.

The Incubation of Normal Accidents
Barry Turner may have been the first to identify the importance of precipitating factors in hazardous scenarios. According to Turner (1978, p. 33), "man-made disasters" are not "bolts from the blue." Rather, when systematically analyzed, disaster scenarios reveal that early signs are often misinterpreted and even ignored (ibid., p. 86). Turner stressed that disasters, especially those that are outcomes of human agency, should be addressed as socio-technical problems that involve social, organizational, and technical processes that through interaction can produce devastating consequences. Specifically, Turner noted that hazardous contexts often have a starting point steeped in culturally or institutionally accepted norms of safety. Obviously, better understanding this early "incubation period" is important to avoiding the onset of a disaster and the subsequent salvage and readjustment.

Focusing on organizational attributes that precede industrial crises, Charles Perrow exposed the catastrophic potential that is "normal" to systems that are both interactively complex (exhibiting the potential for unpredictable synergy between parts) and tightly coupled (having small margins of error). *Interactively complex* systems have the potential for interaction between components that are exogenous to expected (e.g., normal) production sequences (Perrow 1984, pp. 72–84). Simply put, such nonlinear interactions between system parts are unpredictable and hence difficult or impossible to remedy. On the other hand, *tightly coupled systems* (in contrast with loosely coupled ones) are characterized by greater time dependence in their production sequences (ibid., pp. 93–100). Owing to the "tightness" of their sequencing, these systems have little "slack" (small margins of error) to allow for unpredictable interaction between components (a "normal" even if unexpected occurrence, according to Perrow). Examples of such systems include nuclear power facilities, chemical and petroleum refineries, and aircraft operations. In contrast with much of the previous research on high-risk technologies, Perrow locates the risks associated with these complex technical systems in organizational configurations rather than in operator error, and he focuses on the systematic creation of hazards rather than dealing exclusively with their results.

Lee Clarke (1989), studying organizational response to PCB contamination of a federal office building in Binghamton, New York, shed light on the role of organizations in assessing, mitigating, and accepting risks. In this case, goals were especially ambiguous for the agencies involved because no scientifically legitimate safe level of exposure had been established for PCBs, because no consensus existed on how the monitoring of individuals exposed to PCBs should proceed (if at all), and because no institutional or experimental method existed for remedying or cleaning up the contamination. Moreover, because the event was atypical, normal jurisdictions, spheres of responsibility, and decision-making heuristics were vague at best, and in some cases they promoted organizational conflict. According to Clarke, such a scenario cannot be understood either through psychological perceptions of risk nor through an isolated assessment of each individual organization's standard operating procedures. Because the context of risk identification, risk monitoring, and risk remedy involved multiple organizations, it required inter-organizational negotiation and exchange. Elaborating on Cohen, March, and Olsen's (1972) "garbage can model of organizational choice," Clarke focused on how inter-organizational decisions concerning novel risks and ambiguous circumstances led authorities to apply known, even if ineffective, solutions to a problem they did not understand. The "garbage can" theory directs attention away from conventionally conceived notions of organizations as rational problem solvers. It asserts that in problematic situations organizations look like a trash can full of people, solutions, agendas, recognized problems, and choices that are only loosely associated. About organizational response, Clarke (1989, p. 30) says that, "rather than goals directing policy" (as a rationalist argument would assume), "actions were taken and then justifications later sought for them."[28] The outcome was that organizations applied previously effective solutions, but to the wrong problem, thereby initially increasing the risks of contamination.

Diane Vaughan (1996), in her research on the 1986 *Challenger* accident, also explored organizational misdiagnosis of risk. Vaughan held that the manufacture of the Space Shuttle was characterized by simultaneously occurring "production cultures" and "cultures of production," which conjoined to promote organizational maladjustment to risk(s). She found that this setting promoted organizationally derived normalized deviance that explained how unacceptable risks were accepted by NASA engineering work groups and how, somewhat inevitably in retrospect, the catastrophic explosion of the Shuttle ensued. Vaughan's conception of *production culture* captures how "standard operating procedures" or production routines become tacit and assumed through micro-interaction and repetition and, in so doing, tie the members of the work group together intimately. Over time, engineering groups working on the *Challenger* developed norms, values, and procedures that laid the basis for a consensual outlook. The shared production culture, a de facto production paradigm, "developed incrementally, the product of learning by doing." It was "based on operating standards consisting of numerous ad hoc judgments and assumptions that they developed in daily engineering practice" (Vaughan 1996, pp. 394–395).

Cultures of production, in contrast, are definitions of situations that inform sense making in common directions. As cultural scripts filling the gaps between official routines and actual practice, cultures of production represent a common engineering cosmology that propagates conformity among engineering work groups through an overarching culture of production that makes "risky" assumptions acceptable. Examples of such an engineering culture of production include believing in the legitimacy of bureaucratic authority relations, adhering to standardized rules, and taking for granted assumptions about costs, completion schedule, and safety satisficing (Vaughan 1996, p. 396). Thus, according to Vaughan, the perennial and nagging question of why decent individuals take part in work that has bad outcomes is best addressed in the cultural-organizational nexus: "Culture, structure, and other organizational factors, in combination, may create a worldview that constrains people from acknowledging their work as dirty."[29]

In a notable exception to the "normal accident theory"[30] of Perrow and others, theorists of the "high reliability organizations" (HRO) school focus on how certain organizations avoid accidents. That is, HRO theorists focus on the "high record of reliability" that they observe in many highly complex, tightly coupled, and potentially hazardous contexts that organizations must manage.[31] This is in stark contrast with "normal accident theory," which concentrates on inconsistencies observed, on accidents that have happened (or nearly happened), and on the inability of monitoring systems to predict interactive synergies and human fallibility.

In studying high-technology settings, HRO theorists identify five elements that they associate with decreased chances of accidents: safety as a priority for organizational elites (La Porte and Consolini 1991), effective trial-and-error learning from previous (identified and identifiable) mistakes, decentralized decision making (Wildavsky 1988, pp. 125–147; Weick 1987), redundancy (both technical and in personnel; see Morone and Woodhouse 1986, p. 44), and a culture of reliability that is the outcome of strictly defined rules and of intense socialization and training (Weick 1987, 1993; Weick and Roberts 1993).

Of specific interest to this discussion, Karl Weick (1985, 1987, 1993, 1995), in his research on HROs and hazardous contexts,[32] has emphasized the importance of organizational culture in avoiding accidents such as those that Perrow and others claim are "normal." According to Weick (1985, p. 124), a properly socialized and acculturated organization promotes stability without the rigidity inherent to rule-bound, reactive, and routine bureaucratic forms: ". . . it creates a homogenous set of assumptions and decision premises which, when they are invoked on the local and decentralized bases, preserve coordination and centralization. Most important, when centralization occurs via decision premises and assumptions, compliance occurs without surveillance. This is in sharp contrast to centralization by rules and regulations or centralization by standardization and hierarchy, both of which require high surveillance."

Weick, championing organizational culture as a means of avoiding catastrophe, concedes that such a prescription has its dark side if an organization is not continually vigilant. Becoming inured to changing circumstance, he suggests, is the outcome of ongoing routine, assumptions that are no longer questioned, and the social inertia that results. He argues (1985, p. 119) that people inhabiting organizational settings must recognize that "[organizational] inertia is a complex state," that "reliability is an ongoing accomplishment," and that the emergent character of many organizational contexts entails constant organizational attention to the potential for catastrophe. Similarly, La Porte and Consolini (1991, p. 27), in their research on HROs, have found that "operators and managers in these organizations have learned that there is a type of often minor error that can cascade into major, system wide problems and failures . . . and there is a palpable sense that there are likely to be similar events that cannot be foreseen clearly, that may be beyond imagining." The dark side of such heavily socialized prescriptions will become apparent as we delve into the Guadalupe spill's creation and the response to it. Instituted to avoid mishap, and championed by HRO theorists, safety cultures and emergency protocols are effective only if the hazard to be avoided is known and if those involved seek to avert it. Although these prescriptions are important in designing an organization to avoid acute and recognized outcomes, they do little to dodge scenarios that, in the words of La Porte and Consolini (1991, p. 27), are "beyond imagining" or that when recognized are actively covered up for reasons of organizational and individual viability (Sagan 1993).

Outside of negligence, then, an organization must recognize a problem as worthy of attention in order to actively plan against it. As we will see, Guadalupe's chronic spilling never moved beyond "routine mode" for most of those involved (La Porte and Consolini 1991).

3

A "Secret" Spill: When Routine Work Becomes Criminally Negligent

It is like the [newspaper] article I read. It said the foxes were ruling the hen house. Unocal was basically told, "You oversee yourself." If you make a mistake, turn yourself in. That can't be like that. That's got to stop.
—former worker at Guadalupe oil field, interviewed in 1997

If reactions to a potential threat hinge on a display of immediate and consequential impact, then the diluent leaks at the Guadalupe oil field provided little in the way of motivation to act. In such instances of industrial effluent discharge, the responsibility by law and by regulatory protocol is on the polluter to report it. Whereas traditionally conceived forms of law enforcement are predicated on a belief that lawbreakers are going to do everything within their power to avoid getting caught, the local, state, and federal agencies charged with enforcing environmental regulations depend heavily on the violator to inform them of their transgression(s). This "safeguard" against environmental degradation has been codified as law through a number of state and federal statutes that require the self-reporting of chemical and petroleum releases.[1] When any entity, from a mom-and-pop gas station to a corporation, spills more than a barrel of petroleum, the onus is on that entity to report the spill before it damages a waterway or some other significant resource.[2]

Explaining why the chronic spillage of petroleum at the Guadalupe Dunes continued unabated for 38 years, was concealed by field personnel, was publicly denied by their supervisors and superintendent, and went unacknowledged by Unocal's head offices is not as simple as it may appear. Popular explanations of industrial mishaps (or, as in this case, negligence) frequently involve prosaic explanations that locate blame in an individual or a coterie of individuals who, through carelessness or sheer ignorance, precipitate "accidents" (Vaughan 1999, 1996; Perrow 1984, 1986; Turner 1978). Operator error is a primary focus of official blame in 60–80 percent of Perrow's (1984) accident scenarios. However, through extensive analyses of a range of organizational settings, Perrow (1997, p. 67) concludes that "the story of our time is big organizations and big environmental problems," not isolated individuals creating big problems.[3]

Likewise, operator error does not explain why the leaks at Guadalupe would continue for 38 years without a report of responsibility and without efforts at rectification. First, the chronic leaks cannot be rightly characterized as accidental. An accident is an unforeseen, unplanned, often sudden event that is an outcome of chance. The spillage at Guadalupe was not a chance occurrence. Second, responsibility goes beyond an individual or a few individuals responsible for causing the spill or for keeping it secret. Guadalupe's contamination occurred over such a long period that many different workers operated the pumps, piping systems, and storage facilities and held supervisory positions. At any time, scores of workers knew of and did nothing about the repeated leaks and spills. After 38 years of chronic spillage, only two would break the silence; the rest would have to be subpoenaed and compelled to testify. Neither whistleblower would tell for decades, until the leaks had accumulated reaching enormous proportions. The first whistleblower called the authorities in 1990, after working at the field for 12 years. The second called in 1992, as the spill was being tentatively investigated by the authorities. The second whistleblower was on disability leave when he heard a local supervisor lie about the extent of the spill on the local news. He called the authorities because the "underground and surface contamination at the Unocal field has been bothering him for a long time and he could no longer remain quiet regarding Unocal's deliberate coverup" (White 1993).

Claiming that "operator error" does not account for 38 years of spillage is not the same as saying that field personnel were not responsible for making bad decisions or that they were not knowingly negligent, as they obviously were by the 1980s. It does, however, point out that there was something more complicated at work than a few bad employees spilling millions of gallons of petroleum. Similar to Turner's (1978) observations across the cases of man-made disaster he analyzed, it reveals a more systematic, ordinary character to leaks at the field. It also exposes how vulnerable we are to industrial excess with the weak system of industrial self-regulation (also referred to as "self-audit") that is currently in place.

Another equally limited yet somewhat antithetical explanation of why Unocal would choose to deny its responsibility is to envision the corporation as an undifferentiated mass, devoid of individuals, that saw single-minded in its pursuit of profits. The logic out of which this understanding emerges considers Unocal's profit motivations and organizational structure as synonymous and deterministic. According to this line of reasoning (referred to in the literature as the "amoral calculator thesis"[4]), Unocal compared the costs of stopping the leaks and reporting them against the benefits of not stopping them and not reporting them. Having weighed the difference, Unocal decided to hide the spillage in order to avoid the direct costs of changing its method of transporting diluent or of cleaning up what had already been spilled. In short, this explanation would contend that Unocal found it less expensive, and hence expedient, to allow petroleum to leak, hoping that no one would notice. Although profit underlay the field's production routines (and thus underlay the chronic leaks), it does not, as the sole determinant, explain complications that were due to an interaction of complex organization, occupational culture, and individual agency (both at the inter-organizational production-unit level and on the part of individual managers and field workers). Put another way: If corporations, such as Unocal, are entirely profit-motivated "optimizers," then they should have ceased spilling this valuable petroleum product. It is plausible to assume that, had corporate managers and accountants at Unocal's head offices in Los Angeles and Connecticut been aware of the continuous leaks, they would have called for some kind of action, if only to avoid losing profits, even if they refrained from officially reporting it. (There is no way to know from outside the corporation if Unocal headquarters knew about the leaks and spills; this was never established in the court case, and my informants were not in a position to report on it.)

In any case, Unocal's operations on the central coast of California are notoriously dirty. Whether Unocal headquarters knew of the mess being created at Guadalupe and allowed it or whether headquarters did not know of it but implicitly tolerated it by not enforcing environmental standards does not really matter. In either case, Unocal behaved as if oil leaks and spills were not of concern to the corporation, and this was mirrored at the local level. This poor business practice will cost Unocal hundreds of millions of dollars.

According to Einstein (1994, p. B1), the Chevron Corporation—California's largest oil producer—stopped using diluent in its California fields in the early 1970s because of its monetary worth: "[Chevron] found more efficient ways of recovering oil with . . . steam, and the diluent was a valuable product" . . . that Chevron uses to make things like fuel for jets and locomotives."

Even at the low market price of a barrel of Central California petroleum ($1.5–$4 a barrel, at 1980 prices), the conservatively estimated 8.5 million–20 million gallons of lost diluent would have been worth $12.8 million–$80 million.[5] The price of fixing the leaks was much lower. Furthermore, in view of the public-relations debacle that followed the spill's discovery, the legal fees, the costs associated with the cleanup, and the cessation of petroleum extraction at the field, the costs of the spill have indeed become significant. In 1994, Unocal paid $1.5 million in civil penalties for the Guadalupe spill and another $2 million for emergency remediation (*Oil and Gas Journal* 1994; Unocal Corporation 1994a). In 1996, Unocal Corporation dedicated $77 million in reserves to identify the extent of the contamination and to begin the cleanup of the Guadalupe field and another regionally contaminated site.[6] As of November 1998, Unocal had spent $40 million on emergency remedial actions at Guadalupe, and another $43.8 million after the civil case concerning the spill was settled.[7] It is likely that the costs of lost petroleum, the $83.8 million Unocal has spent or is obligated to spend, and the estimated $100–$200 million it will spend in the coming years to clean the site far exceed the costs associated with replacing an antiquated piping and management system (Cone 1998; Finucane 1998).[8] Logically, in order to protect the real and potential profits of the corporation, Unocal should have focused on stemming the flow of petroleum into Guadalupe's sand dunes.

Similar to explanations rooted solely in operator error, the model of the corporation as a profit machine is simplistic and does not explain the circumstances. The fact that organizational directions and individual preferences are the interactive outcomes of one another—in ways that are not under the complete control of the organization's owner, its CEO, or its stockholders—complicates the understanding how decisions are made and ultimately who is to blame when something goes awry. Perrow (1997, p. 69) comments that organizations, although they are ultimately the tools of their owners, are "quite recalcitrant tools." He continues:

They are living systems of humans with interests and attachments of their own, and employees are often able to command special knowledge or strategic leverage to protect those interests and attachments. Their interests may defeat the benign logic of the master who sees how he can profit from good practices. Interests grow up around bad as well as good practices in organizations . . . These groups—superiors, and the technically challenged managers and workers—have what might be called structural interests—interests created by their positions in the structure of the organization. These interests are not necessarily in harmonious concert with those of the masters. An analysis of organizations and the environment cannot ignore them.

In view of the potential for heavy fines, criminal charges, and closure of the field (all of which now have occurred), one wonders why self-reporting did not happen. The qualities that describe the social organization of the oil field—managerial hierarchies, a formally defined system of seniority, internal promotion, and strong social cohesion—appear benign and indeed are common to many workplaces (Jacques 1990). However, it is in and through such a seemingly innocuous organizational environment that one of the largest petroleum spills in US history was created, left unacknowledged, and for some time actively denied. As Vaughan (1999, p. 274) has noted, "much organizational deviance is a routine by-product of characteristics of the system themselves"—of a system instituted to achieve an end efficiently and unproblematically. Vaughan refers to this as the "dark side of organizations."

The intersection of three elements helps to explain why the managers of the Guadalupe field remained unconcerned with chronic leaks for so long and why they concealed it from government regulators once they recognized the threat it posed. First, similar to Turner's (1978) articulation of a "starting point," the organizational routines at the field provided a cognitive template that engendered basic assumptions about how work was to be done and where a worker's responsibilities lay within this larger framework. Second, and intertwined with the first element, the occupational culture at the field lent field routines social stability, thereby playing an important cohesive role in the field's social organization. This cultural framework was a part of the field workers' cosmology—a cosmology that was reiterated through workers' participation in the work group (Weick 1993). This common outlook provided workers with a set of taken-for-granted understandings that, though unwritten, guided field workers in making "appropriate" decisions when confronted with choices (Weick and Roberts 1993).[9] Third, inter-organizational isolation, founded in a loose relationship with other branches of Unocal and with regulatory agencies, insulated the Guadalupe field from the "outside" and gave management, production routines, and cultural affinity more power over individual workers. That is, the Guadalupe field's physical and organizational isolation from other production units, from corporate headquarters, and from regulatory authorities amplified the local managers' power over field personnel and also engendered an in-group mentality among workers that further diminished the likelihood that the spill would be reported expeditiously.

Self-serving motivations, especially on the part of local field management, were another big part of what eventually became an active denial and coverup of the spill. Yet my analysis does not suggest either pure self-serving "evil" intent or "oversocialization." Rather, it suggests that decent people can participate in dirty work.[10] The ensuing analysis fleshes out these dynamics, showing how and why knowledge about the leakage was kept "in-group" until the "daylighting" of petroleum reached critical proportions and the first "rogue" employee finally broke the silence in 1990. Only with the conjunction of these events did news of the spill make its way to authorities and to the local media.

By juxtaposing the premise on which self-report legislation is based (i.e., that compliance is ensured by the threat of steep fines) with the organizational and cultural qualities that characterized oil work at the Guadalupe field, this chapter shows just how flawed such policy is.[11] Because the current system of industrial monitoring is premised on the assumption that industrial dischargers will regulate their own activities (Yeager 1993; Wolf 1988; Skillern 1981), it is crucial to understand why they do not do so in certain cases.

Intra-Organizational Complicity

The part complex organization plays in bounding the rationality of group members is well represented in organizational theory. Bounded rationality conveys the idea that no person, committee, or research team, even with unlimited resources, can attend to the infinite number of stimuli presented in any social or organizational context. The best that is achieved in any circumstance is "bounded rationality"—decision making that is sensitive to specific cues and, by extension, insensitive to others. Specifically, "bounded rationality" is meant as a theoretical substitution for the assumed "omniscience" that characterizes the relatively prescriptive neo-classical and microeconomic theories of decision making. In Herbert Simon's (1955, 1956) original formulation, intendedly rational behavior is always constrained by limitations rooted in computational shortcomings and in organizational structure. Simon's descriptive theory of bounded rationality suggested that human beings develop decision making procedures that are sensible (i.e., rational) in view of the constraints (boundaries) they face, even though they may not appear sensible if those constraints are not discerned (March 1978, p. 587; March 1956). In short: Through socialization processes, organizational contexts heavily influence the grounds for "rational conduct" (Merton 1940; March and Simon 1958; Weick and Roberts 1993; Jepperson 1991; Powell and DiMaggio 1991).

Typically, organizational efficiency is realized through the application and promulgation of extensive, predictable, and consistent programs to guide the actions of those who inhabit them, so that they achieve some specific or designated function (Simon 1947; Turner 1978; Perrow 1986). Work systems, such as that at the Guadalupe oil field, regiment how individuals relate to one another and to their environment (Merton 1940; Simon 1956; March and Simon 1958; Weick 1979; Vaughan 1996). However, as Weber observed around 1900,[12] rationally striving to achieve an end can also lead to irrational outcomes.[13] This insight will help us understand the creation of the Guadalupe spill. By not only following the orders of managers but also internalizing and carrying out local organizational mandates, workers at the field eventually put themselves out of work and subjected the larger organization (the Unocal Corporation) to a significant public-relations debacle and a large financial cost.

Thus, in order to understand why chronic petroleum leaks were unsurprising to personnel at the field, it is important to address the structure of oil work. Over time, as petroleum extraction was rationalized from its early and rather loosely organized entrepreneurial roots, it evolved into an elaborate system of work.[14] Institutional assumptions were integrated into the consciousness of workers through daily enactment, becoming tacit with time. In this way, work routines fostered assumptions about what was important and what was not important. For many years, leaking oil was considered unimportant.[15]

We also must distinguish between data and information. According to Clarke (1993a), information is data transformed to aid in choice. Clarke emphasizes that, although "raw data" and "information" may be objectively the same, it is through social structure and culture that meaning is imputed to data, thus converting the data into information: "The processes through which data are amassed and made sense of determine whether they become relevant information or are dismissed (or simply tolerated) as background noise." (ibid., p. 315) In short, organizational frameworks provide a context that confers meaning, turning data into usable information on the basis of which appropriate choices can be made. Perrow (1997, p. 72) agrees, explaining the power of organizations to delimit reality as follows: ". . . organizations have a profound (and quite essential) ability to shape beliefs and values of their employees, to construct worlds that only outsiders find unrealistic . . . a successful organization is one that consistently and pervasively reinforces beliefs in its core 'phantasies.'"[16]

Nevertheless, it would be improper to conceive of organizational structures as residing "outside" the individual worker. Though founded in programmatic settings, routinized organizational processes are also reified and perpetuated in individual cognition through consensual validation (Weick and Roberts 1993; Weick 1979; Nelson and Winter 1982). Organizational systems act as a "grammar in the sense that it is a systematic account of some rules and conventions by which sets of interlocked behaviors are assembled to form social processes that are intelligible to actors" (Weick 1979, p. 3).[17] In large measure, information is not data until it appears consistent with positions already held (Clarke 1993a). Moreover, preferences do not necessarily precede action (as typically conceived) in means-ends notions of rationality. Rather, they often lend retrospective sensibility to what has already occurred (Cohen, March, and Olsen 1972; March and Olsen 1979).

Functionally, then, organizational actors (a.k.a. workers) are directed to perform on the basis of organizational scripts (institutional rules, protocols, programs, and organizational routines). These are internalized and cognitively reiterated when enacted (Goffman 1959; Berger and Luckman 1966). Simply put, "a way of seeing is always also a way of not seeing" (Turner 1978, p. 49).

A "Starting Point"

In its early stages (from 1953 until 1978 or 1979), the leakage at the Guadalupe field was not troubling, nor was there anyone to whom to report it. Because it was part of routine fieldwork, it received little attention. According to those who read the meters that tracked the coming and going of the diluent, "many times there were little leaks; that was just normal" (field worker, telephone interview, 1996). A worker quoted in a local newspaper went so far as to say that "diluent loss was a way of life at the Guadalupe oil field" (Friesen 1993). Dumping hundreds of gallons of diluent into the dunes, as long as it was done a gallon at a time, was an ordinary part of production. This is not a great leap of reason; oil work obviously involves oil. Until the 1970s, Unocal sprayed the dunes with crude oil to keep them from shifting and thus to make field maintenance and transportation easier.[18] If spraying crude oil over the dunes was unproblematic, why would diluent leaks, which were largely invisible as soon as they hit the sand, be unsettling? Although workers mention that they became alarmed in the 1980s when puddling diluent periodically appeared as small ponds on the surface of the dunes, the chronic leaks themselves evoked little attention. In brief, at Guadalupe the normalcy of spilling oil of all kinds (crude oil, lubricants, and diluent) worked to blunt perceptions of the leaks as problematic. The leaks were an expected part of a day in the life of an oil worker. According to the *Telegram-Tribune* (Greene 1993b): "A backhoe [operator] at the field . . . for 12 years . . . cited 'an apparent lack of concern about the immediate repair of leaks or the detection of leaks.' Diluent lines would not be replaced unless they had leaked a number of times or were a 'serious maintenance problem. . . .' Although workers checked meters on the pipelines and looked for leaks if there was a discrepancy, often a problem wasn't detected until the stuff flowed to the surface' said . . . a field mechanic."

By both historical and contemporary accounts, oil spills have long been a common occurrence in oil-field operations.[19] This seems to have been especially the case at fields operated by Union Oil. But this does not help us understand why, once field personnel recognized the spillage as a significant problem, they denied it and failed to report it (as specified by state and federal law) for 10 years or more. As a first step in understanding why workers failed to report their spill to the authorities once it had "tipped" toward becoming a grievous problem requires us to attend to the vocabulary, the structure, and the enactment of work and how these factors not only molded workers' perceptions of the leaks but also kept them from reporting outside their local work group.

The "Company Line," 1978–1993

Organizationally, oil work at the Guadalupe field was arranged, like work in many traditional industrial settings, around a hierarchical seniority system. Recruitment and promotion were internally derived, meaning the field workers relied on their immediate foremen and supervisors for instruction, guidance, and ultimately, future chances at success (promotion, salary increase, choice of shifts, and so forth). According to a field worker (telephone interview, 1997): "They don't bring people in from the outside. Most of the time it was in house." Hierarchy, as an organizational strategy through which authority is exercised, responds to an organizational need to guide subordinates in their work. Hierarchy invests authority in the supervisor as the embodiment of organizational imperatives. Nonetheless, the "efficiency" of strictly programming duties and defining powers does not come without a tradeoff. The strength of an organization's hierarchy and the power those in positions of authority wield, though immediately "efficient" for getting tasks done, has other costs in the long run—costs that potentially make strict hierarchies less attractive as an organizing principle. For instance, increases in the efficiency of decision making (in a strict hierarchy) can decrease the general sense of personal responsibility. Hierarchy, as an organizational structure, relieves individuals of responsibility for making decisions concerning what is right or wrong socially (substantive rationality, in Weberian terms) and promotes insulation from a sense of moral culpability (Vaughan 1996; Hummel 1987). This organizational structure helps to explain Unocal employees' silence about the Guadalupe spill after it was recognized as a threat. Even when the leaks began to look more like a bona fide spill, the rank-and-file workers were insulated from reporting it themselves by their position within the field's hierarchy and their immediate responsibilities. Reporting outside the work group was management's domain.

A Unocal field worker I interviewed in 1997 articulated his experience of the change from a normal to a problematic spill as follows: "You come up and you see a clamp [on a] diluent [pipe]line. It is leaking. You tighten it up, you change it, you . . . fix it and it . . . has made a puddle. That is not something you would turn in. When it went into the ocean . . . and you see the waves break and they weren't breaking white [but] brown water, there is a problem. [That happened] sometime in the 1980s. . . . We all knew right then . . . we had some kind of problem. Well, we all kind of estimated it could be rather large considering that this field had been here so long before we ever got there." Corroborating this worker's impressions, another worker quoted in the *Telegram-Tribune* (Greene 1993b) remembered finding large concentrations of diluent that were no longer the "leaks" that had created them but looked more like a typical "oil spill": "In 1980 a large puddle of diluent that had saturated the sand and bubbled up to fill a spot 5 to 10 feet wide. . . ." He told investigators that he and his co-workers realized at the time there were problems with the diluent system "even though management seemed to ignore the problems."

By this time, the problems brought on by "normal operating procedures" were obvious and destructive. This became especially apparent in the mid 1980s, when accelerated spillage periodically slowed oil production at the field (Greene 1992b, 1993a; Rice 1994). Yet, instead of self-reporting the spillage, the field workers turned to denial and secrecy. At this point, in a rather Orwellian fashion, supervisors at the field engaged in doublespeak, making claims that no one believed but everyone accepted, for their own reasons as well as for collective reasons. For example, one field worker, interviewed in 1997, commented as follows on the "company line" that supervisors and foremen took about the spillage and on the knowing looks and shrugs their explanations elicited from those who worked the pipelines, pumps, and storage units and who everyday saw diluent lost into the dunes: "One day down at 5X [a well head and extraction site] I took a shovel and dug down by where [the ocean] came up on the sand, and diluent started coming out. . . . We talked [to the workers who had witnessed it] and we reported it to Union's foreman. They said 'If anybody asks, tell them to talk to the superintendent.' At a safety meeting the superintendent says 'Ah, well, a boat probably went by and emptied its bilges.' We all looked at each other and [shrugged]; you know, come on." In the same interview, this field worker commented on the increasing intensity of the leaks at the field as the pipeline infrastructure deteriorated: "Once it hit that point [in the mid 1980s], it was hundreds of barrels a day—a couple of hundred a day being missed, and you know only one barrel out on the surface is allowable. . . . We used to talk about [the spillage]. Everybody knew. The fact that we would lose 200 barrels a day, you would think that if we lost 200 barrels a day and were having a problem. . . . But it was business as usual." At this time, workers realized their jobs were increasingly in jeopardy, since the entire field might be shut down. Some workers wanted no part of the newly defined problem; others wondered what the consequences of discovery would be. Moreover, some acted as if "they didn't know anything, and didn't want to know anything, all they wanted to do was keep their jobs"; others admitted they "had a problem and wondered when, and if, it would come out."

Once the leaks were perceived as threatening, they imbued local organizational routines with a form of "institutional schizophrenia": When conflicts arise between public ideology (i.e., it is against the law to spill) and institutional action (i.e., continued spilling), individuals are likely to separate the two and to shift between the two "levels," employing "a form of double-think" that "is a major result of means-ends conflicts" (Bensman and Gerver 1963, p. 597). At Guadalupe, workers distanced themselves from the continued spillage by separating it from a right-or-wrong conceptualization. By the 1980s, workers knew that spilling was wrong, but only if they were responsible for it. In view of the hierarchy in the field, they were not responsible; their managers were.

The hierarchical insulation from responsibility thus helped to keep workers who watched diluent spill into the dunes from feeling obligated to do something about it. When relieved of making decisions, people tend to cede their personal responsibility to those who are in control (Milgram 1974; Asch 1951).

The field's hierarchy had five major levels.[20] A new worker began as a utility man, then worked his way up to pumper and then to field mechanic. If able, with long enough tenure at the field he could become a foreman. Over the foremen were the field supervisors, who headed operations at specific fields; over them was a superintendent who oversaw Guadalupe and another oil operation in the area. Parallel to the field's organizational hierarchy was a formally defined seniority system that distributed authority within hierarchically defined categories. In a seniority-based system, newer workers rely on those with more seniority—who have already done their jobs and learned their lessons—to manage them and to inform them on how to perform field duties (Glazer 1987; Sherman 1987; Bensman and Gerver 1963). One worker put it this way: "I had to learn about equipment that I had never used. I was assigned to be with other people that had been there for a lot of years. . . . They would break you in. That is how you would learn." (field worker, interviewed in 1997)

The lower-ranking workers at the field left reporting of the spillage to the foremen, who were to notify the field supervisor. In turn, the supervisor was to inform the regional superintendent, who then was to inform environmental experts within the corporation, corporate executives, and the appropriate government agencies. Those on the lower rungs, particularly utility men, pumpers, mechanics, and vacuum truck drivers, were not in a position to report the spill outside this hierarchy—at least, not without breaking rank and risking sanction (Glazer 1987; Elliston et al. 1985).

When the second of two whistleblowers—"disgusted" by the lies he had heard a Unocal supervisor tell on television[21]—called the authorities, he informed them that the spill extended into the interior of the field and was being covered up by managers. California Fish and Game wardens then raided Unocal's local offices and found plume maps and well logs delineating petroleum spills that local field managers had not reported. Those maps indicated that local managers had been keeping nominal track of the spills since the 1980s (Paddock 1994a,b; Rice 1994; White 1993; Finucane 1992; Greene 1992b, 1993a,b). Claims that managers kept the spillage quiet were corroborated by other field personnel I interviewed, as well as by those quoted in the press. For example, an environmental compliance specialist at Unocal's local offices related that he was "sicken[ed] and disheartened . . . because [the spill] could have been prevented and should have been cleaned up way before now" (Rice 1994, p. 31). He left Unocal—"still unaware of what had happened at Guadalupe" (ibid.)—after local managers chastised him for spending $200,000 on the cleanup of another oil field site close to Guadalupe. Furthermore, he and another environmental specialist were quoted as saying "they were kept at arms length" and that "Unocal managers . . . consider[ed] the environmental compliance department a "necessary evil" (ibid.).

Moreover, in the mid 1980s workers were reporting it to their foremen when the leaking diluent had accumulated and reached sizable proportions, but that information remained with the foremen and their immediate supervisors. In 1986, according to a California Fish and Game warden (interviewed in 1997) who had deposed Guadalupe oil-field employees in conjunction with the state's investigation, workers suggested to foremen that pumps be installed to gather lost diluent; the foremen rejected this proposition because they "didn't want anyone to know that the diluent was there to be seen." In the words of a field worker interviewed in 1997: "I think [information about the spill] went to one spot. . . . It did not get reported higher up. . . . I was always under the assumption that they tried to keep it in their little area and keep it a little secret." Another worker, interviewed in 1996, remarked: "It went up the chain of command. It went to the foreman and he reported it to his boss. It stayed local."

The culpability of all those at the field, but especially the superintendent, supervisors, and field foremen, coupled with the field's organizational characteristics, meant that explicit knowledge concerning the scope and scale of the leaks stayed inside the local operation. "Each field is its separate own little field," said a field worker interviewed in 1997. "We were kind of out in the middle of nowhere. So once we reported to our superior [a field foreman], then he has to report it to the field supervisor, who has to report to the regional superintendent, who then reports it to Los Angeles. Some where along the line I think it stopped. I think that it stopped with the field supervisor." This field worker was describing a *loosely coupled* organizational arrangement—one with organizational units that are "somehow attached, but [whose] attachment may be circumscribed, infrequent, weak in its mutual affects, unimportant, and/or slow to respond"[22] (Weick 1976, p. 3). In this case, the slack that existed at the local field between workers and between workers and managers and the loose organizational coupling that existed between the local field and corporate offices (including environmental divisions) were reflected in the technical division of responsibilities, in the authorities of office, and in the expectations placed on each. A great deal of flexibility existed between these units as long as certain goals were met. In this case, petroleum continued to be produced and sent out at an acceptable rate. In view of the local field's autonomy and field personnel's collective interest in remaining a viable production unit, not telling outsiders about the spill made a great deal of sense.

The mainline workers' passive resistance and (especially) management's active resistance to reporting the spill in the 1980s is also understandable in view of their accountability for its long-term occurrence and the punitive consequences they faced. Vertical organization and internal promotion meant that many of the managers, over the long term, held a great deal of retroactive responsibility for the spillage, insofar as promotion entailed working one's way up through the same sequential set of job titles and responsibilities once held by those currently in positions of power. Once a worker made it into management, he was implicated in all those years of spillage. Thus, of all those at the field, the foremen, the supervisors, and the superintendent had the least motivation to tell about the spill once they realized how big it was. Not only did they have the longest history with the spillage; they also stood to lose the most. When interviewed in 1997, a field worker explained why foremen were not motivated to report the spillage, grounding his assertion in their long-term knowledge of (and their hence accountability for) the chronic leaks: "His motivation not to tell? Well, if he came out and told, how is he going to explain all the years in between? What are they going to say: 'Oh, well, we decided to tell you now, but let's forget those other years'? Plus, there was a time that the (field supervisor) was a foreman at Guadalupe, before I came there. I know that during that time, they also had diluent problems."

The long-term nature of Guadalupe spill made it especially problematic for all those who worked the field for any length of time. Liability for it was diffuse—indeed, organization wide. For those in the lower echelons, going outside the proscribed line of command to report the spill created triple jeopardy: Not only would they risk being personally associated with an organizational offense; they also would have been informing on co-workers and endangering their careers by implicating their superiors. One does not succeed within a vertically organized work setting by "ratting out" one's superiors or co-workers.[23] Fear of social and organizational reprisal was evident in my discussions with field personnel, in California wardens' accounts of their interactions with subpoenaed field personnel, and in local newspapers' stories such as Greene 1993b: "Current employees contacted for this story were surprised and dismayed their names would become public because of what they told the state investigators. They worried about their superiors and co-workers at Unocal finding out."[24]

Workers at the Guadalupe field did not want to go over the head of their field foreman, their supervisor, or their superintendent. A field worker, interview by telephone in 1997, said: "There is somebody above you and someone above them and someone above them. One thing that you don't want to do is break the chain of command . . . that causes friction." Informing might have affected how many hours of work one received, one's chances of promotion, and ultimately whether or not one would keep a well-paying job.[25]

Negative sanction, however, is not the only condition encouraged by such an organizational framework. The promise of promotion also reinforces allegiance to the established hierarchy (March and Simon 1958; Sagan 1993; Collinson 1999). At Guadalupe, internal mobility motivated conformity to organizational rules (written and unwritten); it also fostered strong identification by individual workers with the field as a social organization. In addition to coercing those who might have informed to keep quiet (Garfinkel 1956), this dynamic implicitly reinforced keeping quiet through praise and advancement (Glazer 1987; Bensman and Gerver 1963).

A Culture of Silence

Local managerial power and organizational routines did not wholly determine behavior at the field. The normative framework that prevailed there was also attributable to the subculture of oil-field work and to individual workers' agency. Turner (1971) defines industrial subcultures as a complex aggregations of norms, job roles, social definitions, explanatory frameworks, and moral injunctions, the existence of which are crucial for the maintenance of continuity in human activities within industrial endeavor. In the case of worker agency, while both worker roles and social membership in the field's culture strongly influenced individual choice, individuals also acted according to their own interpretations of those structures based in personal needs. Field workers applied corporate directives in ways that were not necessarily to the benefit of larger corporate interests. Ironically, individual agency is also observable in the whistleblower's decision to report the leaks after 10 years or more of silence.

Surprisingly, in the winter of 1988 Unocal's field managers reported oil on the beach to the National Response Center (NRC), the US Environmental Protection Agency (EPA), and the California Regional Water Quality Control Board. However, as the following excerpt from the 1988 spill report makes clear, those accounts were not admissions of responsibility; rather, field management claimed that the oil had not originated at their operation: "Samples sent to [Unocal's] research lab shows source was not from the Monterey Formation [a petroleum reservoir under the Guadalupe Dunes] or injection fluid [diluent] used in the Guadalupe field. Thus, sampling [by Unocal] shows source is not Union's operation." (California Regional Water Quality Control Board Spill Report 1988)

To understand more fully why workers kept quiet about a spill they knew was patently illegal while field managers covered it up and lied to authorities about its origins, we must look beyond matters of hierarchy and seniority. We must look at individual motivations and at the social glue that bound workers to their work group. In short, we must look at the dominant social milieu at the oil field in order to see how social relations between workers played into the initial normalization of the spill and how they reinforced the intra-organizational conditions that discouraged self-reporting. Taken separately, both structural and cultural explanations would predict that self-reporting was unlikely; together, they make self-reporting appear a dubious regulatory strategy.

Vaughan's contribution to our understanding of organizational conformity sheds light on why, when spillage at the Guadalupe field became a significant problem, the leaks continued and why field personnel actively denied them. According to Vaughan (1996, p. 409), one of the most significant lessons that her work on the *Challenger* accident reveals is how "the formation of a worldview . . . affects the interpretation of information in organizations." Vaughan goes on to relate how in the *Challenger* case this entailed environmental and organizational contingencies that generated "pre-rational" orientations that shaped collective sense making. This worldview worked to normalize signals of potential danger, fostering the reiteration of mistakes that had negative consequences. Furthermore, Vaughan adds, it is not only in the original development of group goals that deviance is normalized. It is also in their reiteration that the incremental expansion of normative boundaries takes place. This incremental expansion not only habitualizes the participants to deviant events; it also increases their tolerance for greater levels of deviation. This neatly captures both the initial tendency of personnel at the Guadalupe oil field to see the releases of diluent as unproblematic and to accept ever-larger releases of petroleum as a normal part of the job. (By the late 1980s, hundreds of barrels at a time were being lost, some in diffuse leaks and some in more concentrated spill events.) Vaughan's comments also illustrate a nexus of circumstances and social elements. Organizational structures affected field culture, but equally as important, field culture lent work stability to routines (more precise, to the routines' ramifications: the spill and the silence about it). The field's cultural milieu thus played an important cohesive role in normalizing deviance at the field, providing strength and "inertia" to the field's social organization.[26]

Social Ties and Field Secrets

In general, oil work is noted for its strong fraternal work culture (Romo 2000; Collinson 1999; Davidson 1988; Quam-Wickham 1994; White 1962; Hasalm 1972). The cohesion that has characterized the setting through time is in part an outgrowth of the work. According to Nancy Quam-Wickham, who has documented California's oil industry and the production culture that drove it, "solidarity between oil workers grew out of a distinctive set of conditions," and "the work process demanded team-work among members of any individual work unit: producing oil was a synergistic process" (Quam-Wickham 1994, p. 26). This characterization of the historical traditions of oil-field work, certainly applies to the Guadalupe field.

Workers at Guadalupe inherited and developed a set of norms and beliefs about what were and were not appropriate in-group behaviors. This is a normal part of group unity. Moreover, that this unity led to the coverup of an ongoing petroleum spill becomes more understandable (even if socially inexcusable) when we address the threat it posed to each individual at the field and to the local outfit as a whole. According to March and Simon (1958), socialization and the persistence of group goals reflect two socio-cognitive mechanisms. The first of these resides in the ontology of the individual decision maker and is the outcome of the selective perceptions sponsored by group goals. As group imperatives are internalized, individuals become the gatekeepers of their own focused attentions. The second mechanism is an outcome of intra-group communications. The vast bulk of what we know is based not on firsthand knowledge but rather on second-hand and third-hand accounts we receive from others in our reference group(s). Thus, social settings produce frames of reference that act as perceptual filters, consonance among first-, second-, and third-hand accounts of the same phenomena reinforce individual perceptions, and overall persistence of "collective mind" (Weick and Roberts 1993; Vaughan 1996; Asch 1951; Janis 1982).

As we have seen, the kind of trouble that awareness of the spill might unleash silenced the Guadalupe field's management. However, pressure to keep the spill a secret, based in a de facto culture of silence, was not only found in management; it was also observable in how field workers reacted when they found out that one of them had called the authorities. (Again, the first admission came in an anonymous telephone call to state officials in

February of 1990.) When interviewed in 1997, the field worker who initially blew the whistle related being overheard by the field office's secretary and described the secretary's reaction to his phone call as follows:

I got on the phone in the office. I say [to the health department official], "Okay, I'll talk to you later," and I hear his click, and I'm still on the phone, and I hear another click. The secretary eavesdropped and heard my conversation. She came in, and she started yelling at me, "What are you doing! We will all lose our jobs!" And I said, "Not if we didn't do anything! If it isn't ours, why would we lose our jobs? We are not going to lose our jobs!" We knew [about the spill]. But I never thought it would come to the point where they would shut everything down. What I thought would happen is they would isolate the problem and go on producing.

Though it may seem odd that a whistleblower would use a company phone to notify the authorities, a comment he made clarifies why he would feel confident doing so. It also sheds light on his personal struggle in regard to rectifying the spill and his years of participation in its creation.[27] Even the whistleblower did not want to believe that an organization of which he was a member was responsible for an enormous oil spill. Commenting on his "deep-down inside" hopes, the whistleblower, when interviewed in 1997, made it clear that he had been aware that if the spill was huge it meant big trouble for him, for Unocal, and for the environment: "I heard rumors that a sewer system line had broken in Guadalupe and there was sewage. At that point, . . . I did not want it to be ours. I wanted [the managers] to be telling us the truth, and that we could go on [working at the field]. . . . I was [thinking] 'Please don't be Unocal.' . . . But deep down inside I knew. . . . I got tired of seeing it in the water. . . . What is going to happen . . . if they just continue [to spill]? What [are] the long-range effects?"

Individuals, in protecting themselves from association with the spill, also collectively shielded the organization from harm, at least in the short term. The threats of a shutdown of the field and a loss of jobs and the social pressure to remain silent kept workers from reporting the spill. (Once the spill was "discovered" by regulators, Unocal's corporate headquarters did shut the field down, and all the workers were either transferred or laid off.)

Moreover, breaking with one's peers and eliciting an out-of-group admission about what was (initially at least) a "normal" part of production was also unlikely for a set of more socially relevant reasons. Even once the spill had accumulated and became noticeable, reporting it would mean informing on co-workers and facing their opinions.[28] Once his identity became known at the site, the whistleblower was ostracized by many of his fellow workers. In the following quote from a 1997 interview, he relates how quickly word of his telephone conversation with the authorities traveled and how it affected his relationships with co-workers: ". . . the minute I left the office, the secretary talked to another guy on the [oil field], and as I drove by I saw them talking. She was talking a hundred miles an hour. From that time on, this guy wouldn't talk to me; a lot of the other people did not talk to me. Some did, some were cool about it. . . . 'Hey, this is not your fault; this is something that was going to come out eventually.' A week later, all of a sudden, it was all over TV and everything. I came to work, and they wanted to know 'Did you call the media?!'"

Social sanction is especially severe for individuals who spend most of their week, and over the long term most of their life, with fellow workers (Glazer 1987; March and Simon 1958).[29]

Inter-Organizational Location as Amplification

In conjunction with the organizational location of workers relative to one another and to management and with the culture of silence that character-ized the field, the Guadalupe field's structural isolation from outside inter-ference (both physically and organizationally from regulatory authorities and Unocal's head offices) and the corporate incentives worked against self-reporting. Like many other corporations, Unocal was not a monolithic undifferentiated body with a single objective or universally shared knowl-edge. In organizational form, Unocal consisted of loosely coupled upstream corporate offices, production units, and downstream refinery and vending segments. Insulation from outside interference amplified the power that field routines and the local production culture had over individual percep-tions and over field workers' choices.

Because the Guadalupe field was largely autonomous from its head offices, its day-to-day domestic affairs were largely internal. A report of an incident had to go to the top before making its way to outside authorities. Because the information stopped in the field's chain of command, it never made it out of the field, where action could be taken to stem it. This is not a claim that Unocal headquarters could not have known about the spill if they had wanted to investigate it. The argument forwarded here is more passive: Headquarters was interested only in specific information from Unocal's extraction divisions, and this information tended to consist of production quotas rather than of information as to whether environmental matters were being addressed. Again, Unocal, as a corporation, seemed to care little about how local operations performed their production as long as the fields continued to produce profits.

Historically, the Guadalupe operation's relative isolation from outside interference seems an ordinary aspect of oil production. For example, the historian Gerald White (1962, p. 522) characterized oil-field superintendents and foremen circa 1900 as "little kings" who "did pretty much as they pleased, so long as they conducted their operations efficiently." Whatever efficiency has meant (generally, low overhead and high oil production), it has not, for most of the industry's history, included environmental concerns (Pratt 1978, 1980; Molotch 1970).

Had efficiency included not wasting diluent, a case could be made that the loss of diluent into the dunes would have been a sign to those on the outside that something was amiss. In this instance, Unocal's head offices may have taken a more active interest if dollars were being lost. Had hundreds of thousands of gallons of refined petroleum product been purchased from an external source and subsequently lost, it would seem expensive and hard to cover up. But spilling was considered a part of producing oil at Unocal's operations, and it also was rather normal for others in the industry. Furthermore, it was considered largely an internal affair. The diluent used at the site beginning in the early 1950s originated at Unocal's refinery situated at the edge of the Guadalupe Dunes, literally a part of the Guadalupe field's production infrastructure. Oil extracted from the Guadalupe field was piped to the Nipomo refinery for initial separation. Diluent, as a by-product of this refining process, was then pumped back to Guadalupe for use. At Guadalupe, diluent was stored at a number of tank farms; from there it was transferred via pipeline to individual extraction wells. If production was consistent, lost diluent would not be missed, especially in view of the normality of spilling and the shoddy records that were being kept (because the price of refining was internal). Losing diluent cost the local operator little (at least, relative to getting caught or facing the prospects of personally reporting it), as long as crude oil was being produced at the expected rate. On the other hand, if the field supervisor reported the spills (which had "tipped" toward the obvious in the 1980s) he would have known that he had a big monetary and criminal problem on his hands. It would have tarnished his personal record, reflected badly on Unocal's image as a whole, potentially shut down the local operation, and presented the possibility of criminal prosecution.[30] What is more, the potential fines for having not reported the spill are significant.

Two examples of the penalties associated with pollution of this sort illustrate the predicament that field managers confronted when deciding whether to report the spill. The federal Clean Water Act specifies that violators can be fined between $5000 and $50,000 a day per violation for being "knowingly" negligent.[31] Estimating the potential fines involved for this single act would require starting with the date of the amendment's passage (in 1973) and calculating daily fines up to 1990 (when Unocal ceased using diluent at the field). The estimate ranges from $31,025,000 to $310,250,000. Likewise, under California's Proposition 65 (a citizen-sponsored "right to know" act passed in 1987) Unocal was also liable for not reporting its release of petroleum into local river and ocean waters frequented by recreationalists. Proposition 65 caps fines against violators at $2,500 per person per exposure day. These are but two examples.

Moreover, the field supervisor and superintendent personally stood to lose thousands of dollars in potential bonuses that were paid for meeting corporate expectations. Field supervisors received incentives in the form of commendations, quick advancement, and end-of-the-year bonuses for keeping production costs down and petroleum yields high. High production costs would have resulted from capital outlays for such items as Guadalupe's pipeline infrastructure. Much as in the system that prevailed in the Soviet Union into the 1980s, costs were "hidden" by a reward system that recognized only production goals and the accompanying steady income stream. Thus, the primary goal was keeping production high, not worrying about diluent costs that (at least on paper) were trivial, being locally internalized. According to a Unocal supervisor interviewed in 1997 for this research, why it took 38 years for the spill to be reported by field managers was rather easy to understand: "Unocal [did not report the spill] to the public because local managers received financial incentives to keep costs low. The corporate culture of the production outfits saw spills as a normal part of their routine." Although this was not the only reason that local Unocal managers would continue to spill, it certainly provided a strong incentive not to report it or stop the leakage at the field once it had become organizationally ominous. Only negative personal and organizational repercussions would have resulted if local managers reported the spill. As a latent product of the pressures articulated thus far, spilling and not reporting makes a great deal of sense from the production side of the equation.

Conclusion(s): A Normal and Criminal Spill

What is new in the preceding analysis that has not been touched on by other examinations of industrial behavior? It falls easily within the mainstream as defined by organizational theorists, especially the theorists of normal accident theory (Vaughan 1996; Sagan 1993; Perrow 1984). Yet some ramifications of the foregoing analysis are not explicitly entertained by theorists of normal accidents and by other organizational theorists who do not typically address how creeping problems are created, left unacknowledged, and eventually actively denied. In short, important features distinguish the Guadalupe scenario from the scenarios addressed by previous empirical investigations. First, the circumstances and the setting of the Guadalupe spill depart from the technically complex, tightly coupled, and nonlinear interactions that have been the primary interest of much "accident research" (Perrow 1984, 1986; Morone and Woodhouse 1986; Weick 1993; Weick and Roberts 1993; Sagan 1993; Vaughan 1996, 1999; Sagan 1993; La Porte and Consolini 1991; Wildavsky 1988; Roberts 1982). The loose organizational coupling that characterized intra-field and inter-field relations, the unsophisticated technologies in use, the linear qualities of pipeline transport and metering, and the long duration of the spill all set it apart from the organizational setting previously examined. Second, accident theory has typically focused on settings that hold the potential for acute and catastrophic outcomes. This chapter focused on the gradual development of a problem that, with time, became enormous. Third, the analysis is unique for its attention to the "stages" that constituted the "incubation" of this major environmental disaster (Turner 1978). Identifying the additional phases the spill went through with a focus on social and organizational processes deepens Turner's notion of a "starting point." It reveals a side of industrial crisis that is seldom investigated and exposed.

The purpose of this chapter has been to convey how banal social and organizational characteristics in a low-technology setting can lead to alarming results of a crescive and insidious type. It is out of such commonplace structures, coupled with equally unremarkable yet incrementally accumulating trouble, that destructive consequences emerge in the long term. This is in contrast to Clarke's assertion (1989, p. 158) that "all important accidents, by definition, produce social disruptions." In some cases they do not, which is precisely why they need more research attention. Integrating an organizational analysis of the dynamics of an oil field with an understanding of power and personal interest, and by relating that analysis and that understanding to disaster theory and to the literature on organizational deviance and that on whistleblowing allows me to propose an answer to the question of why the accumulating trouble at Guadalupe was initially unrecognized and eventually actively denied—an answer exactly the opposite to the desired one, and an answer that contradicts the logic of cooperative compliance as the basis for industrial regulation.

As outlined in the preceding pages, the leaking petroleum at the Guadalupe field, as a meaningful event, went through a number of stages based on normal operating procedures, organizational structures, the social milieu, and the inter-organizational location of the field. In the first instance, the spill's long gestation was characterized by an acceptance of what had already leaked and of what continued to escape. Accommodation to these conditions led field personnel to overlook and for sometime misinterpret signs of petroleum accumulation. This is in line with Turner's (1978) observations concerning the norms of safety that frequently underlie the incubation periods that preceded catastrophic events. To borrow from Perrow (1984): The spill truly was "normal," if not so "accidental." However, the normalization of the chronic leaks was not static. As time passed, not only did individuals at the field accept what had already spilled; as routine leakage intensified as a result of infrastructural deterioration, workers accustomed themselves to these increases and to the lies of their supervisors, thereby allowing the amount considered "normal" to grow with time.

When it had "tipped" toward an obvious problem in the 1980s, even for field workers used to the sight and smell of oil, the protracted spill had promoted organization-wide complicity that had canceled any chance of self-reporting. Had the spill been sudden and dramatic, the regulatory response could have been more immediate, individual operator error could have been more easily applied, a select few individuals could have been purged from the corporation, an apology could have been issued, and (if required) a fine could have been paid.

However, because of the duration over which the spill occurred, multiple workers had had direct experience of it, had failed to report or stop it, and hence had some responsibility for it. Those who had started work as utility men at the field and had stayed for many years moved into foremen and supervisory roles, and by the 1980s they knew of the oil under the ground and the continued leakage into the Santa Maria River and the Pacific Ocean. They also knew that everyone else was as guilty as they were. Everyone's hands were dirty with spilled oil; everyone's guilt (and personal stakes) ensured that no one would tell (Becker 1963; Glazer 1987; Sherman 1987). Like Bensman and Gerver's (1963) observations concerning the deviant use of the "tap" screw in aviation manufacture, the oil spill at Guadalupe, once it clearly was a threat, worked to tie field workers together on the basis of their group membership, their participation in the spill, and the consequences that might result from its discovery.[32]

Those at the bottom of the field's hierarchy who did not have long-term participation in spilling and not reporting had little motivation to report the spill for another set of related reasons. First, they all chanced losing their jobs if the field were to be shut down. Second, reporting the spill would have meant informing on their superiors. Third, reporting the spill would have marginalized any of them socially from his co-workers (Glazer 1987). Reporting is not simply a personal matter; it is also a moral accusation against other co-workers because they did not report it themselves (ibid.; Garfinkel 1956). The first of these points (and in spirit the second and third) was well articulated by the secretary when she screamed at the whistle-blower: "What are you doing! We will all lose our jobs!" As was touched on earlier, she was correct. Soon after the spill was reported, the field was shut down and all those that worked there were transferred, demoted, or (as in the whistleblowers' case) fired.

Moreover, having such an occurrence discovered on one's watch is something no one wants as a part of his work history (Collinson 1999; Sagan 1993). And what happens if one takes it upon himself to remedy the situation by going as far as reporting it? Even with current whistleblower legislation, the whistleblower stands a good chance of being marked a snitch, which makes the old work environment unlivable at best (i.e., the whistleblower is ostracized) and off limits at worst (i.e., he is fired) (Elliston et al. 1985; Perrucci et al. 1980; De Maria and Jan 1996). And work in other outfits for other companies is also off limits: No one hires a whistleblower (Glazer 1987; Westin 1981).

Coupled with production incentives, inter-organizational location (i.e., isolation) only worked to increase the odds against the reporting of any mishap, because resistance was located at the top of the local field—the only place where reporting was formally acceptable. In many ways, local management's resistance to acting proactively to stem the spill, once they realized that it was a "losing course of action" (i.e., a threat to their continued organizational viability), appears similar to what Staw (1981, p. 577) has referred to as "rising commitment in escalating situations." Such situations are characterized as those in which sunk costs (which can be economic, psychological, social, organizational, or all these at once) affect whether individuals or organizations cease a "questionable" line of behavior or escalate their commitment to their current dubious course of action. According to Staw and Ross, what is especially interesting is that in situations characterized by "slow and irregular declines" (as opposed to acute and immediate shocks) "persistence" in losing courses of action may grow stronger as a result of organizational inertia (Staw and Ross 1989, pp. 209–210). This seems especially close, for obvious reasons, to the situation that unfolded at Guadalupe.

In brief, organization-sponsored complicity, the culture of silence, and the inter-organizational isolation of the field combined to make reporting of the spill Guadalupe improbable until the accumulation of diluent had gotten so bad that neither insiders (field personnel) nor outsiders (regulators) could fail to recognize it, a society of environmentalists was there to be concerned about it, and the insistent local media were eager to report on it. These are all factors that society can ill afford to either count on or wait for.

Theoretically, then, the analysis in this chapter builds on research that attends to the precursors of man-made disasters, looking at the conditions under which the long incubation and slow accumulation of dangerous circumstances occur (Turner 1978; Vaughan 1996). Specifically, the foregoing analysis of the causes of and the responses to the Guadalupe spill by oil-field personnel adds to organizational theory concerning "how aspects . . . normally associated with the bright side of organizations are systematically related to the dark side" (Vaughan 1999, p. 292).[33] This chapter has also stressed the importance of combining a view of power and interests with organizational and cultural qualities as integral to a disastrous outcome.

An important caveat must be stated here, insofar as this chapter has focused more on social and cultural factors (the banality of the leaks, the shared social construction of what constituted "oil work," social cohesion between workers, and the organizational location of the local outfit) than on why a few managers at an enormous multinational oil company would (criminally) resist reporting their spill and would silence those who might have reported the leaks and spills: The predominantly social, cultural, and structural explanations I have put forth are powerful in part because of the nature of industrial regulation in the United States, where it has been left to corporate actors to report their own excesses. There is little interdiction, investigation, or active following up of problems by government authorities until a situation is so dire that a coverup is impossible to sustain. Thus, outside of personal motivation on the part of a worker, a foreman, or a supervisor to report a leak, there is little (aside from morals) that would impel anyone in a company to do so. And that is a slippery slope that takes us back inside the social dynamics that characterized the normative and cognitive institutions that characterized Unocal's local field operations.

Thus, the influence that current self-reporting requirements had on the events at Guadalupe are worth noting. Though the spill was obviously not a product of the current regulatory environment, the largely reactive and punitive system, combined with a self-reporting requirement, engendered its own set of unanticipated consequences. At least in Guadalupe's case, such environmental measures did not encourage the kind of cooperation they were supposed to produce. Once the spill had become troubling, not reporting it became even more expedient for the local outfit, in the face of—and in some measure because of—the punitive regulatory environment. Perhaps punishment coupled with self-reporting represents the worst of both worlds. This lays bare a fatal flaw of current regulatory systems that combat industrial excess: They implicitly assume that organizations will report on themselves and, on the other hand, that the "masters" of such complex organizations have the power to make their workers tell on themselves. This flies in the face of what the present chapter has brought to light. The very foundations for producing and reproducing an organization (both at the manifest organizational level and at the latent cultural level) would predict they would not do so. Thus, even if we were sure that large organizations would self-report their violations, Local work groups and individual workers have the very real capacity to act on their own, despite overarching organizational goals (Perrow 1997; March and Simon 1958).

4

The Agency Beat: Waiting for a "Tanker on the Rocks"

Interviewer: Who initiates and decides whether it is emergency or remedial action?

Regulator: Well, interestingly it was a negotiated process.

Interviewer: Negotiate whether it is an emergency or not?

Regulator: That's what we did. Because it wasn't real clear if it was an emergency or not—*it wasn't a tanker on the rocks.*

After decades of sloppy maintenance, coverup, and the spilling of millions of gallons of refined petroleum product, it took repeated community complaints of foul odors at the beach and a whistleblower to get the enforcement, judicial, and remedial balls rolling.[1] At first glance, responsibility for this spill would seem to fall unquestionably and squarely on Unocal. That, however, addresses only the source of the petroleum, not why the regulatory systems currently in place to monitor such industrial activities would not react to the leaks until they had become the nation's largest petroleum spill. The following pages recount the elements that help elucidate why agency response came so late and why it had to be premised on an emergency before the "silent spill" could be addressed.

Unocal began using diluent as far back as 1952 (Stormont 1956). Questions as to why regulators responded in 1990 and not in the 1960s, in the 1970s, or in the early 1980s led me to investigate the policy agenda (and the public agenda) of those earlier periods. Obviously, conceptions of pollution, and of its negation (a "clean environment"), had changed. Thus, in addition to the initial normalization of the leaks and then the internal complicity that characterized the oil field that kept them a virtual secret, there was the general lack of societal response to such industrial practices until the late 1960s or the early 1970s. Although it would be improper to characterize "pollution" as a post-1970s construction, its suppression as an item for the public agenda helped to keep it out of the civic domain for many years (Mazur 1998; Weale 1992; Pratt 1978, 1980; Enloe 1975). Before the 1970s, beach walkers, fishermen, and surfers who smelled petroleum or saw dead foliage and might have wanted to report it had no recourse except to tell field supervisors or foremen of their impressions. That still fails to account for 20 years during which Unocal's spilling at the field intensified. For example, in 1985, in 17 days, 11,298 gallons of diluent (equal to 269 barrels) were spilled at the field (Greene 1993a). Thus, it was on the basis of a perceptual change that the leaks at the Guadalupe Dunes were re-defined as an "oil spill" rather than as "oil leaks," and it was on that basis that they eventually became an item on the agenda of the government agen-cies charged with public safety (Crenson 1971).

Though not without vacillation, collective concern about environmental health (broadly construed) has increased with time (Beck 1992a; Weale 1992; Lash et al. 1996; Gottlieb 1993; Yeager 1993; Nash 1989). Never-theless, pollution itself and pollution-abatement measures are in need of qualification. While regulatory systems have increasingly addressed pollu-tion, institutional response remains reactive and motivated by a narrow set of concerns. Remedial systems are largely indifferent to chronic pollution scenarios. They wait for moments of "extreme density"—moments when a potential threat has become critical (Erikson 1994). Moreover, regula-tions and the regulators charged with enforcement tend to focus on partic-ular media (air and water as the primaries), often at the expense of others (groundwater and, until quite recently, soil) (Weale 1992). In the case of the Guadalupe Dunes, recognition and reaction would come only after the leaks fulfilled a set of definitions based on their intensity and the resources they affected.

Obviously, responding to the Guadalupe spill meant seeing it in a certain light. Early on, as with the oil-field workers, the spillage held a to-be-expected status with state and federal regulators or response would have ensued. According to those who regulated the area, the sight and the smell of petroleum at the dunes and at the beach were familiar to them. "What do you expect?" said a California Fish and Game warden interviewed in 1997. "They produce oil there!" Over time, signs of the spillage became less ordinary and more compelling. Nevertheless, while new legislation and institutional actors began to take note of the diluent on the beach, they did not aggressively pursue a remedy. At this point, the Guadalupe Dunes spill did not fulfill criteria that would have made it an "emergency" petroleum release, nor was it a benign occurrence. It was perhaps a problem, but not a pressing one. Commenting on this, a California field biologist who had been involved with the initial investigation of the spill (in 1990) attributed the low priority of the event for state and federal regulators to its lack of "sex appeal": "This was not an *Exxon Valdez*. I think that is one reason the state was slow to respond. Not that there weren't a number of state employees who were concerned about it and working to try and figure out what went wrong and what should be done about it. But, as an agency, Fish and Game and the Coast Guard weren't organized to do anything about non-sexy, non-public sorts of chronic pollution."

Once the leaks had reached undeniably threatening proportions, in the winter of 1993–94, these same administrators began to broach a novel problem for which they had no comprehensive or institutionalized solution on which to base their reactions. The spill crossed jurisdictions and called on agencies to coordinate in ways that had not been done before. For instance, it was not clear just how extensive the diluent contamination was or the kinds of dangers it presented. Moreover, proposing and implementing solutions proved equally difficult and divisive. Solutions involved developing ways of coordinating responses and applying remedies that had to be acceptable to eighteen local, state, and federal agencies, as well as a community that increasingly became angry about the spillage. In the end, addressing the problem entailed defining priorities and applying solutions to a case of contamination that no one understood and for which there was no remedy. It looked more like a long-term management project than the "usual" emergency cleanup.

The following pages outline three conceptual stages through which the regulatory response to the spill proceeded: the stage of ordinary leaks, the stage of a problematic emergency, and the stage of long-term management of pervasive contamination. We begin with a brief glance at federal environmental legislation (and the socio-historical context in which it is embedded) that concerns pollution, pollution abatement, and industrial oversight.[2] This is to provide an institutional thermometer of sorts: the re-definition of these regulatory priorities is a product of changing legislative agendas. Although passing legislation did not directly result in the discovery of the Guadalupe spill, the change in priorities at the federal and state level brought the spillage increasingly into the purview of regulators. From this history, we move into an analysis of the reaction the Guadalupe spill initially received, focusing on how regulatory response reflected ambiguities inherent to official conceptions of industrial crises and environmental threat. Focus will initially be on why regulators were originally unresponsive to the spill and secondly why their response required reconceptualizing it as an emergency, in order for a reaction to occur. Finally, through an analysis of the formal intuitional processes surrounding site characterization—whereby the extent of the dunes' contamination was ascertained and solution proposed—significant yet largely implicit preferences are exposed. I refer to this as *negotiating ecology* to capture how priorities were defined (for instance, aesthetic, health, and evaluation of future use) as well as the struggles that emerge from what are often conflicting valuations of the environment (Shrader-Frechette and McCoy 1993; Shrader-Frechette 1994; Longino 1990; Weale 1992). The irony is that ecology, the interconnection of a biome's organic and inorganic elements, is non-negotiable. Negotiating environmental tradeoffs is a strictly human endeavor that entails hierarchy, incomplete knowledge and understanding, values, and vested interests among a long and often unarticulated list of variables that are part of the science and politics of decision making.

From Gushers to Spills: The Policy Agenda and Pollution Control

In the previous chapter I outlined the bounded rationality that prevailed at the oil field. A similar statement can be made about those farther afield who regulate industrial enterprise. Policy prescriptions, agency missions, and their relation to the industry they regulate play a crucial role in defining what is, and is not, problematic. Federal and state environmental regulations are of relatively recent vintage. Until 1948 (5 years before Unocal purchased exclusive rights to the Guadalupe field's oil reserves), the only environmental acts of any significance were the federal government's Refuse Act of 1899 (which forbade the dumping of waste in navigable waters), the Public Health Act of 1912 (initiated to stem disease by keeping waterways clear of debris associated with bacteria), and the Oil Pollution Act of 1924 (which banned oil discharge in coastal waters). The last of these specifically targeted petroleum discharges because the law in place at the time (the Rivers and Harbor Act of 1899) had proved ineffective in the face of new conditions not common at the time of its enactment. Rapid industry expansion after World War I led to a crisis in the quality of coastal waters that served to focus regional and national attentions on such degradation (Pratt 1978). Although the Oil Pollution Act of 1924 was important in establishing the legal basis for pollution control, it had little if any impact on industry because of its limited enforcement powers.

In 1948, with the Water Pollution Control Act, the federal government took its first clear steps toward a genuine environmental agenda concerning issues of pollution abatement. The Water Pollution Control Act mainly acted as a conduit for the states to receive research funds and technical assistance when such needs arose. Although (much like the earlier Oil Pollution Act of 1924) the act did not amount to much as far as enforcement was concerned, it did lend official recognition to a growing national distress over water pollution. Similarly, the Air Pollution Control Act of 1955, though also largely a token gesture, worked to give air pollution and related issues a forum they had not had (Colella 1981). Beyond symbolism, these acts set the stage for local and regional regulation of industrial discharge, which eventually would effect industrial practices. Through federal financial assistance to state governments, the development of a public body of knowledge concerning pollution, and growth in the number of experts, the public sector, for the first time, had a means of critically addressing issues of pollution and pollution control (Pratt 1978). Behind regulatory legislation that followed on the heels of these early acts, the federal government increasingly moved toward establishing itself as the primary enforcement body (Gottlieb 1993; Colella 1981). What had been left to local authorities and private industrial initiative through the 1950s became the domain of federal regulators in the 1960s. The first of the federally enforced environmental initiatives was the Clean Air Act of 1963, which imposed national standards and gave federal authorities the power to mandate compliance with them.

Toward the end the 1960s, the general mood concerning environmental quality soured as events such as the burning the Cuyahoga River in Cleveland, the Santa Barbara oil spill, and the extensive degradation of the Great Lakes (Lake Erie in particular) gained widespread public attention (Gottlieb 1993; Molotch and Lester 1975; Colella 1981). Beginning with the passage of the National Environmental Policy Act in January of 1970, the 1970s witnessed a plethora of environmental acts and amendments whose intent was to set standards, outline enforcement powers, and lay down assessment protocols for the protection of the environment. The flurry of activity marked a substantial change in how US regulatory institutions viewed and approached pollution. In the case of the National Environmental Policy Act , the burden of proof would be placed on the producers to ensure that the quality of the human and natural environment would not be compromised by their activities (Bradely 1996). For example, important amendments to the Clean Water Act[3] and to the Air Pollution Control Act paid increasing attention to protecting air and water resources from the disposal practices of heavy industry and from effluents associated with diffuse source points such as automobiles. Included with the promulgation of these new environmental acts and provisions was the creation of new agencies and governmental divisions to handle environmental degradation and species loss.[4] It is important to note the role of the 1969 Santa Barbara oil spill in the reauthorization of the Clean Water act in 1972, for the spill was less than 100 miles south of San Luis Obispo County and left an indelible mark on local sentiment toward oil. Though the 1972 amendment was ostensibly passed to ensure cleaner water, its motivation was the "the horror of spreading oil from platforms off the coast of Santa Barbara" (Burger 1997, p. 97).[5]

The paucity of regulation attending to land-based disposal was partially remedied in 1976, when Congress passed the Resource and Recovery Act and the Toxic Substance Control Act. According to Szasz (1994, p. 11): "As late as the mid 1970s, the disposal of these often highly toxic materials was almost totally unregulated. Federal clean air and water statutes provided some 'authority over the incineration, water, and ocean disposal of certain hazardous wastes'; fourteen other laws dealt with various aspects of hazardous waste management 'in a peripheral manner'.[6] No federal law regulated land disposal, by far the most prevalent form of industrial waste disposal."

Emerging from rising concern about the proliferation of potentially hazardous chemicals and industrial wastes, the Resource and Recovery Act and the Toxic Substance Control Act attempted to address these concerns by regulating their use, treatment, and disposal. The speed with which fear of toxicity progressed from social and political neglect to widespread dread, a social movement, and legislative attention is truly remarkable (Crenson 1971; Colella 1981; Gottlieb 1993; Szasz 1994). Before the publication of Rachel Carson's book *Silent Spring* (1962), popular anxiety about environmental toxicity was marginal. Indeed, chemicals were thought to herald a new age of health and longevity (Tenner 1998; Humphrey and Buttel 1982). However, by the late 1970s and into the 1980s (by which time diluent spillage at the dunes was chronic and pervasive) fear of environmental toxicity had spurred an unprecedented social movement (Szasz 1994). Were the air and the water clean? What if one lived near a garbage dump, a waste incinerator, or a nuclear facility? And (of great political significance in view of the increasing publicity afforded toxic contamination events in New York, Michigan, Missouri, and Massachusetts) was the average middle class neighborhood truly safe from such insidious threats? In particular, the events at Love Canal, New York, charged issues of toxicity with a great deal of urgency. The plight of Love Canal's residents, whose homes had been built in proximity to a toxic waste disposal site, left many Americans in fear for their own health (Mazur 1998; Levine 1982; Gibbs 1982).

Amid such fears, the federal government rushed to implement the Comprehensive Emergency Response, Compensation, and Liability Act and to amend and reauthorize Superfund legislation (Szasz 1994; Gottlieb 1993). The final version of the Superfund Reauthorization Act included "public right to know" provisions, federally funded technical assistance in cases of toxic contamination, and emergency planning protocols.

Beginning in 1975, efforts to create a comprehensive federal oil spill liability, compensation, and response bill intensified based on a handful of spill events (Birkland 1998). However, it would be another 10 years before Congress would agree on a wider approach to the threat posed by petroleum spills. Part of this delay was attributable to congressional attempts to integrate oil spill policy with concurrent and emerging hazardous waste legislation (ibid.). State resistance to a federally sponsored and controlled response system that would preempt state law and jurisdiction was the other concern that foiled attempts to systematize oil spill response and liability (Jones 1989).

Legislation concerning oil spill response and liability was in disarray at the time of the 1989 *Exxon Valdez* accident. Until 1990, the statutes that directly addressed oil spills were the federal Waterways Safety Act of 1972 and the federal Water Pollution Control Act of 1973. The *Exxon Valdez* spill provided Congress the impetus to pass stringent oil spill legislation. The outcome, the Oil Pollution Act of 1990, amended the Water Pollution Control Act (e.g., the Clean Water Act).

Similarly, at the state level, no act existed that directly addressed emergency response to marine spills or gave a single agency jurisdictional preference. Within a short time of the passage in Congress of the Oil Pollution Act of 1990 and of the Huntington Beach tanker spill, California would also pass a similar and complimentary bill, the Lempert-Keene-Seastrand Oil Prevention and Response Act, to improve emergency response at the state level. In both cases, this new legislation gave responsibility for all vessel-related and marine-related spills to specified agencies. With the federal Oil Pollution Act of 1990, the Coast Guard was charged with managing the Oil Spill Liability Trust Fund and coordinating emergency response.

The Oil Pollution Act of 1990, in order to address hazardous substance discharge, called for the installation of a three-tier National Response System, mandating the formation of a coordinating committee made up of national, state, regional, and local governments and/or industry and the responsible party. When a spill occurs, the National Response Center must be notified. A unified command system must then be set up, through which all decisions concerning coordination, cleanup, and so forth are to be made. (See figure 4.1.) The responsible party (in this case, the Unocal Corporation) holds the responsibility for cleaning up the spill; however,

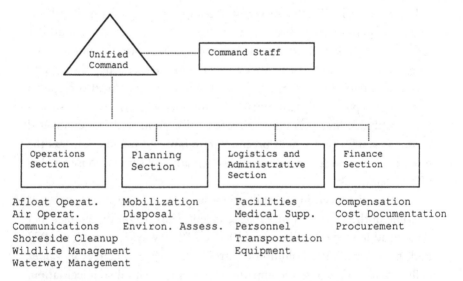

Figure 4.1
Unified command structure called for by Oil Pollution Act of 1990. Source: US Coast Guard 1995a.

the federal on-scene coordinator has the last word on major decisions. When a spill occurs in navigable or coastal waters, federal regulations have put the Coast Guard in the position of supplying an on-scene coordinator, organizing the oil spill committee, and generally being in charge of cleanup operations (Burger 1997).

In California, the state Department of Fish and Game's Oil Spill Prevention and Response unit (OSPR), created under the Lempert-Keene-Seastrand Bill, has the responsibility for directing emergency response in state waters (Elliott 1999).

The Oil Pollution Act of 1990 and the Lempert-Keene-Seastrand Bill systematized emergency response to petroleum releases by giving federal regulators and in some instance state administrators overriding jurisdiction to command and direct emergency action. Specifically, the acts proscribe for emergency marine releases a leadership tripartite to unilaterally take charge (depending on the resources affected) and impel the responsible party to remedy the identified spill. In doing so, the intent was to alleviate confusion and duplication, sponsoring quicker and more effective response in the advent of large-scale and acute petroleum releases—all good things if appropriately applied.[7] Nevertheless, in order to be activated the "emergency release" criterion had to be met, something that proved problematic for Guadalupe's chronic problems.

From Normal Leaks to Favored Pollution Solutions

The ambiguities that the Guadalupe spill presented for activating agency response were echoed throughout interviews with regulators, in agency memos, in court depositions, and in press accounts of the initial response the spill received and in subsequent remedial actions. If the spill was small (a.k.a., an "ordinary leak"), either it was a "non-problem" or it was taken for granted that Unocal would report it (which authorities, in the face of contrary evidence, assumed for quite some time). Anything more than a barrel and less than 10,000 gallons was Unocal's responsibility to report.[8] According to a Regional Water Quality official (Greene 1994): "Our whole system is based on the honor system, based on trust because with the staff we have in San Luis Obispo we have to cover 300 miles. . . . We can't be everywhere and see everything." Corroborating this claim, a local regulator commented: "I'll be frank with you, one of the reasons why this

went on so long without anybody saying anything was with most of the
monitoring systems in place, the monitoring processes, are self-monitoring. Meaning if you are a discharger—you are Unocal—and you spill
something, leak something, it is your responsibility to report it to the
Regional Water Quality Control Board. That's just the way the system is
set up, it is a self-monitoring system. If you decide that you're not going
to tell anyone, no one is going to know." (Project Leader, San Luis Obispo
Planning Department, Energy and Natural Resource Division, interviewed in 1996) If the spill reached threatening proportions (under a very
particular definition of "threat"), appropriate emergency responses were
outlined in the Ports and Waterways Safety Act of 1972 (subsequently
folded into the Port and Tanker Safety Act of 1978, the federal Water
Pollution Control Act of 1973, and later into the federal Oil Pollution Act
of 1990 and California's Lempert-Keene-Seastrand Act of 1990). Yet the
Ports and Waterways Safety Act and the Port and Tanker Safety Act were
not oriented toward policing oil leaks, and later, with the promulgation
of the Oil Pollution Act of 1990 and the Lempert-Keene-Seastrand Act
following the *Exxon Valdez* and Huntington Beach spills, they were oriented almost exclusively toward tanker spills in particular and crude oil
spills in general. In the words of another state official involved with the
spill: "[California's] Oil Spill Prevention and Response unit was established for the *Exxon Valdez* type of spills. We were not originally formed
to work on these long-term remediation projects. So, politically OSPR
stayed involved Guadalupe as somewhat of a touchy subject, because it
has set precedent for us . . . and not just work on the *Exxon Valdez* types
of emergencies." (field biologist with California Department of Fish and
Game, interviewed in 1997)

In view of the preceding legislative agenda and the spectacular industrial crises that motivated it, the reasons why regulators would initially
not acknowledge the Guadalupe spill begin to emerge. Regulators are
indisposed to respond to hazards during the early "incubation stage"
(Turner 1978, p. 84). Later, with the new spill-response legislation, regulators were on the lookout for *Exxon Valdez* type spills at the expense of
other more chronic pollution problems. This is a zero-sum analogy. In
view of the limited attention and resources agencies and individuals have
at their disposal, attention to one priority in this instance translated into
ignoring another.

However, the story does not end with the inability of agencies to act on such crescive troubles. Once problems at the field were recognized after 1990, response was complicated by different agencies having different obligations (and thus priorities), meaning none of them had complete responsibility, but all had interests they were compelled to defend. In view of its close "working relationship" with the petroleum industry, the state Division of Oil, Gas, and Geothermal Resources may be the agency that holds the most responsibility for not initially "discovering" the leakage.[9] By the late 1970s, the San Luis Obispo County Planning Department, in conjunction with the California Coastal Commission, had permit authority over coastal development. Beginning in the mid 1980s, the Regional Water Quality Control Board began overseeing issues that concerned California's water quality. The California Department of Fish and Game oversaw that area's flora and fauna. And finally, the Coast Guard, as of 1990, had jurisdiction over emergency marine releases (Lima 1994). Though their jurisdictions overlapped, they also had what is known in journalism as "beats" (Fishman 1978) or in the institutions literature as organizational domains (Levine and White 1961; Warren et al. 1974). Organizational domain, as articulated by Levine and White (1961, p. 597), is the "specific goals [an organization] wishes to pursue and the functions [an organization] undertakes in order to implement its goals." Their research examined inter-organizational exchanges, finding that effective organizational interaction is contingent upon defining, and agreeing on organizational fields of attention, or what they referred to as domain consensus (Warren et al. 1974). As we will see, the response the Guadalupe spill received cannot be characterized by such a consensus.

Further complicating this potpourri of protocols and jurisdictions (and the bounded rationality that they induced) were changing definitions of the environment, of pollution, of what constituted threatening circumstances, and of how to remedy it. In retrospect, it would seem that the petroleum leaks were self-evident, and hence enforcement as well as remedial reaction should have commenced. This, however, is only the end of a long spill history. In truth, the spill was re-discovered, in a manner of speaking, in 1990 and again in the winter of 1993–94. At this time, the Oil Pollution Act of 1990 and its state-level parallel, the Lempert-Keene-Seastrand Bill, were in place, making oil spills a new priority for federal and state regulators. In other words, California Fish and Game wardens were ready to see the petroleum spill as a problem, and the Coast Guard was anxious to wield new emergency cleanup powers (derivative of the Oil Pollution Act of 1990)—but only if the spill was a certain kind of "disaster."

Organizational Problem Solving under Ambiguous Conditions

Idealized versions of how organizations approach and solve problems draw a picture of complete and valid information passing through the organization, in a predictable way, to those that make decisions at the top of an organizational hierarchy (Clarke 1989; Weick 1976). According to a rational-choice model, the selection of alternatives is based on the attributes the problem displays, organizational constraints, and decision maker preferences embedded in a criterion of economic efficiency (Thompson 1967; Lindblom 1979, 1959; Simon 1955; Etzioni 1967). This view underlies the assumptions of many organizational theorists.[10]

A theory at variance with the neo-classical rational-choice model, and one that is especially useful for making sense of organizational behavior in uncertain and ill-defined circumstances, is the garbage can theory of organizational problem solving. Forwarded by Cohen, March, and Olsen (1972), the theory addresses how organizations function when their goals remain unclear, technologies remain ill-defined,[11] and participation in decision making is fluid and fluctuating.[12] Characterizing organizational problem-solving behavior in such circumstances, Cohen, March, and Olsen (1972, p. 1) remarked that it appears more like "organized anarchy" than an efficient, rational, and orderly progression toward a solution (as, for instance, a more rational and structural Weberian model would predict).[13]

According to Clarke (1989, p. 174), who used the model in his empirical research, the "garbage can" metaphor of organized anarchy is useful because it "provides a way of thinking about organizational forces that differs from the assumed" (often implicitly) rational and deterministic models. In such contexts, "rather than coordinating their behaviors so that an organization moves toward a well organized end, participants often behave in ways that reflect no plan," or at least little understanding of the greater purpose the organization is pursuing (ibid.). "Solved" and "solution," in this context, take on new meanings.

Often, organizationally favored solutions are haphazardly applied to currently emerging problems. Moreover, if a problem does not "fit" the favored solution(s), problems stand a chance of reinterpretation so that they do. In fact, Cohen et al. goes so far as to suggest that problems in complex organizational settings are unhinged from solutions; rational and structural assumptions of a problem-then-solution progression are turned on their head, suggesting instead a solution-then-problem one. In a counterintuitive fashion organizational logic appears to resemble "answers looking for questions" more closely than the reverse (Cohen, March, and Olsen 1972, p. 3). Importantly, the authors add, it is primarily through seemingly coincidental matching that organizational preferences are discovered; they emerge through enactment (ibid., p. 1). Functionally speaking then, problems, solutions, participants, and choice opportunities are conceptually uncoupled and paired together for reasons largely based on the decisions making context: by the participants that happened to be on the scene, by the favored solutions that are advanced, and by the kind of external pressures that are brought to bear on those making the decisions (e.g., the garbage can) (ibid., p. 17).

Nonetheless, some solutions attach to some problems. The question is why, when, and how? According to Scott (1981, p. 274), in ambiguous goal situations, where organizational priorities (preferences) and procedures (technologies) remain unclear, "ambiguous goal statements are often replaced by more specific, proximal, and often procedural goal statements that provide the basis for making decisions and achieving order" and in so doing alleviating organizational doubt. The salience organizational uncertainty holds for decision makers is considerable. Alleviating it becomes as important to the organization as are the "external" problems generally conceived in rational models as initiating organizational action to begin with. That is, the search for a problem to match the organizational solutions is directed inside to remedy doubt and ambiguity as much as it is to address problems that reside external to the organizations interior machinations (Meyer and Rowan 1977).

Although the elements of Cohen, March, and Olsen's model of organized anarchy is generally mirrored in the institutional response the Guadalupe spill received, a wider more robust application of the *intra*-organizational garbage can fits Guadalupe's circumstances more closely. Much of the ambiguity that surrounded initial response and remedial activity involved overlapping and unclear *inter*-agency jurisdictions and relational authority, at times conflicting *inter*-agency preferences and priorities,

ambiguous *inter*-agency response protocols, as well as fluctuating participation of these organizational participants.

Again, Clarke (1989) has demonstrated similar organizational dynamics in his research on organizational reaction to the PCB contamination of
a government office building in Binghamton, New York. To reiterate,
Clarke found that the real strength of Cohen, March, and Olsen's model
lies in explaining how institutions coordinate activities between one
another.[14] Clarke (ibid.,, p. 174) asserts that "a group of organizations . . .
is more like a garbage can (or crowd), having neither an institutionalized
structure to coordinate its members nor a centralized office that issues
orders. . . . Entry and exit in decision-making opportunities are relatively
easy, and there is no hierarchy that clearly delimits authority . . . among
organizations."

Agency response to the Guadalupe spill closely mirrored the general elements described by garbage can theorists (and, more specifically, Clarke's
inter-organizational garbage can). The theories' general direction lends clarity to why government agencies reacted as they did, a response that otherwise appears as chaos or even as complicity in the spill. Specifically, the
decision-making context within which choices were made included an institutionally indistinct problem (as discussed above), bounded organizational
fields of attention, limited and unclear institutional protocols, fluctuating
participation and jurisdictional overlap, and rigid and conflicting inter-
organizational agendas and priorities. These elements played a large role
in the linkages that would occur between elements found in "Guadalupe's
garbage can": problems, solution, participants, and choice opportunities.

Bounded Fields of Attention

Environmental legislation is supposed to enable regulators. Ironically, legislation also limits them by acting as perceptual filters. Early on, before the
slough of environmental edicts were passed in the early 1970s, regulators
did not "see" spilling petroleum as a problem, as industrial pursuits such
as oil extraction were largely viewed as representing work and progress not
pollution (Pratt 1980; Freudenburg and Gramling 1994; Gramling 1996).
In Guadalupe's case, once legislation to stop or mitigate oil spills was put
into place, officials were unable to recognize it or to pursue it aggressively
until it had the trappings of an emergency. An emergency, according to
newly promulgated federal Oil Pollution Act of 1990 and California's

Lempert-Keene-Seastrand Oil Prevention and Response Act, was a spill that posed "a substantial threat to the public health or welfare."[15] Because Guadalupe provided few acute and "substantial" signs that it was a threat to community or environmental health, regulators, in essence, waited for Unocal to report their own excesses.

When signs of the spill began to surface and complaints of odors at the beach became more frequent, the two primary state regulatory agencies, the Department of Fish and Game and the Regional Water Quality Control Board, initially continued to take Unocal's word that they had no problems at the field. According to Wilcox (1994a), California Fish and Game officials "felt Unocal was being pro-active" and were "not going to waste . . . valuable staff time when they are doing everything they can possibly do." Agencies' trusting Unocal may seem odd in retrospect, but that is the letter of the law. "Self-report" defines what is at the heart of current industrial regulatory system (Yeager 1993; Wolf 1988). In view of instances such as those in the preceding testimony, it is clear that self-reporting also defined a cognitive boundary for those that were in place to enforce environmental law (Weick 1993; Vaughan 1996).

The obvious ramification of such legislation is its promotion of passivity on the part of the regulatory authorities by bounding their fields of attention. It is only in hindsight that the regulatory agencies seem rather naive, waiting for Unocal to incriminate themselves. A California Fish and Game warden noted the pinch that the trust he placed in Unocal had put him. As a regulator, he was responsible—a fact not lost on Unocal lawyers when deposing him in court. According to this warden (interviewed in 1997), Unocal lawyers asked why he hadn't investigated and even whether he had gone to "investigation school." "It was getting to the point," he recalled, "where I was getting defensive."

Organizationally, not only were regulators unable to see the spill; once it was recognized, they were in no position to investigate it. The aforementioned warden, commenting on just how much trust the system puts in the polluter to self-report, continued: "You know they [Unocal] are the professionals in the field [of oil]. I am not. I am just a game warden. 'You people [Unocal] are in the oil industry, you tell me what's going on.' I am in a very vulnerable position. I personally went with 'You are telling me the truth, I believe you.' A major mistake on my part." In the same interview, this warden related his frustration with Unocal Lawyer tactics during his deposition. Unocal lawyers used a "not doing his job" tactic to help build their case and have criminal charges dismissed—that is, Unocal lawyers claimed throughout the deposition that he had not done his job properly. Why, they asked in so many words, would an enforcement person expect them to tell on themselves? The following quote[16] excerpts a portion of the testimony:

Defense: Did you have any conversations with [Unocal managers] with respect to the oil on the beach?

Deposed: Yes, sir.

Defense: What did they tell you?

Deposed: . . . that they indeed had checked this and that the oil that I was . . . looking at . . . was not theirs because they had that oil fingerprinted and it did not match anything that they had in their field.

Defense: Did you have any further conversations with [the field managers] on that date?

Deposed: Yes, sir. . . . It was regarding where are we going . . . as far as responsibility goes. . . .

Defense: Then this is the same conversation when they're telling you that it's not their oil, but they are agreeing to see about cleaning it up? Is that correct?

Moreover, in local press the Unocal attorneys were quoted as saying that Fish and Game investigators had "abdicated their responsibilities by sending a fox into a hen house to count the chickens and then blamed them when they emerged with feathers in their mouths." One Unocal attorney compared the investigators to a police officer who would believe the denials of a suspect holding a smoking gun: "What you have here is paralysis by analysis. . . . 'Yep, it's diluent,' but [they] failed to investigate further." (Friesen 1993).

Enacted to fulfill the requirements of a diverse constituency, legislation that institutionalized self-monitoring sought to simultaneously save money, ensure environmental compliance, and be agreeable to US industry (Yeager 1993). In the first instance, state and federal governments could not (and still cannot) afford to finance all the inspectors that would be needed to examine, pro-actively, the nation's industrial enterprises in a search for regulatory compliance. And when they do investigate, recommendations for followup are often ignored for similar fiscal rationales. In another missed opportunity, after the February 1990 beach remediation project a field investigator for California Department of Fish and Game asked his superiors to conduct tests to assess the biological damage that the site might have incurred as a result of diluent leakage at the beach[17]; further limiting what regulators could attend to, "budget constraints" restricted the department's ability to conduct such research, and the call for site bioassays was not pursued (warden, California Department of Fish and Game, interviewed in 1997).[18]

Limited and Unclear Institutional Protocols

Circumstances changed almost overnight with the call of a second whistleblower in June of 1992. Because there was already a tentative investigation of the spill taking place, the call of this whistleblower reinforced the severity of the problem at the dunes for authorities for several reasons. First, it acted as a "self-report" because a field employee, in calling the authorities, was admitting that Unocal had problems. Second, the call acted as the pretext for an "emergency response," pushing the California Department of Fish and Game to acquire a search warrant that allowed wardens to enter Unocal's private property. Third, the whistleblower gave the authorities information on where Unocal's local managers kept documents that depicted some of the field's contamination. In response, the Department of Fish and Game issued warrants and raided Unocal's offices in Orcutt, where records were kept.[19] Some 50,000 documents were confiscated, along with maps depicting more than 200 leaks of more than a barrel—spillage local Unocal managers had claimed to know nothing about. The district attorney filed 28 criminal charges against Unocal and its six local field managers (Friesen 1993).

Yet, at the time, the agencies that had responsibility to interdict, stem, and remedy such pollution events and to enforce the law lacked consensus on just what to do. Once oil spillage had reached an attention threshold (becoming a generally recognized problem worthy of agency time), it presented federal, state, and local regulators with an ambiguous set of attributes. What should they do with continuing petroleum seepage and petroleum soaked dunes and wetlands? Because institutional protocols for such an event were limited and unclear (no comprehensive set of guidelines was in place to direct their actions), their responses to the spill continued to be tentative, intermittent, and delayed. Even basic assumptions about the nature of what was and was not environmentally detrimental were questioned.

Fluctuating Participation and Overlapping Jurisdictions

Adding to the confusion, fluctuating participation on the part of regulators meant that institutional and local knowledge of the spill lacked continuity, further retarding anything resembling a concerted and continuous effort to remedy it. The flux in regulator participation (attributable to bureaucratic rotations, promotions, and staff attrition) occurring over the relatively short period of local attention (10 years) is remarkable. In only four of the eighteen agencies involved with the spill (the four for which this kind of information was available), there were seventeen different individuals ("principals") who would lead their agencies' investigation, monitoring, and remediation programs at the spill site over the period 1989–1996. This accounting, by no means an exhaustive, is based on the memory of informants and on archival documents obtained in the course of this research. The Regional Water Quality Control Board had five different principals. The California Department of Fish and Game, aside from its changing departmental emphasis, had at least six different principals involved, and several others dropped out. At the San Luis Obispo County Planning Department, three different principals played roles in overseeing the spill. At the California Coastal Commission, at least three principals cycled through. Similar rotations were the rule at other agencies. Commenting on the complexity that this adds to a project when the continuity of decision making is repeatedly interrupted, a lead staff with the California Regional Water Quality Control Board, interviewed in 1996, said: ". . . Different personalities . . . and personal biases come in to play. Some people may have much stronger feeling toward protecting the environment, and another person might be pro-business, . . . They might be less strict in dealing with a company like that, or more strict. . . . And, looking back on it, I can say that if you have continuity, I think its better for the project, its better for the environment. I don't think we'll ever have that just because of the dynamics of this office. . . . It's more a reflection of a big organization."

To the flux in personnel must also be added, at least initially, vacillating institutional participation. This made decision making even more complex, revealing how unclear lines of responsibility were for the spill's regulatory oversight. Depending on the requirements of a proposed remedial action or on the resources affected, different agencies abruptly launched into or pulled themselves out of participation. The Coast Guard exemplified this, entering the remedial process suddenly and exiting just as suddenly after their 1994 call for the emergency beach remediation project. Once the Coast Guard recognized the size of the spill, said Lieutenant Cunningham (the federal Incident Commander), "we had no idea of the size [of the leak near the ocean]. . . . Now that we do know, we have to act." The goodness of fit of federally based emergency response protocols was no longer in doubt; the oil had to be stopped at any cost, even at the potential expense of the beach and the dunes (Bunin 1994). Having reacted to the imminent threat of petroleum release, the Coast Guard handed its responsibilities off to the federal Environmental Protection Agency, which had more experience with large-scale contamination. The EPA had resisted involvement early on (1989–1990), claiming it was not in their jurisdiction (lead staff, US Environmental Protection Agency, interviewed in 1997). A San Luis Obispo County planning commissioner, interviewed in 1997, who had made repeated calls to get official response to worsening conditions at the field, articulated this ambiguity over jurisdiction and her frustration with it as follows:

[In 1990] I called the EPA in San Francisco which could not decide whose jurisdiction this was. They believed that because it was West of Highway 1, they did not need to address it.[20] The Coast Guard did not believe this was in their jurisdiction either.[21] At the time, I also contacted the Regional Water Quality Control Board; they took care of the problem in 1990 with the installment of the bentonite wall [at the beach]. [Later] I also called the Coastal Commission, who had only the 1990 information. Then I called the Department of Justice because it was time for some action. Got some at that point. The EPA sent down a representative to determine jurisdictions. At that point [in 1993], the Coast Guard moved in and took command.

The equivocation of the agencies and their staff members stemmed in part from their difficulty fitting the spill into one or another of their organizational domains.

However, the problem the spill presented went deeper than individual organizations' struggling with whether it fell under their jurisdiction. Not only do organizational domains structure manifest functions, such as the protocols and routines of an organization; in a latent and a priori sense, they also structure the consciousness of what is significant. According to Warren et al. (1974, p. 26), "in a very important sense, the domain is 'prior' to the organization, in that there exists in the institutionalized thought structure not only an area of concern and activity that the organization comes to claim, but also a whole definition of social reality which indicates this area as one for which activities are appropriate and which also provides the basic outline of the manner in which the individual . . . organization will carry on activities within its functional field."

Consensus on domain priority across or within organizations lays the basis for common assessment and hence coordination in instances of cross-jurisdictional problems. In this sense, consensus structures which organization should do what, when, and how (ibid.). In the case of the Guadalupe spill, consensus on what constituted a threatening set of circumstances (counter-intuitively, that intermittent leakage was not a hazard) and (after recognition that it was a significant spill) a lack of consensus as to whose problem it was both played big parts in stalling an official response.

Rigid and Conflicting Agendas

Again, the spill would remain administratively ambiguous until the Coast Guard re-interpreted it as falling within its domain. "At that point," the Planning Commissioner quoted above noted, "the Coast Guard moved in and took command." When diluent release became an acknowledged problem after winter storms in 1992, whether it constituted an "emergency release" as stipulated under the Oil Pollution Act of 1990 became a topic of debate. It was not precisely clear under whose purview the spill fell. According to local officials' recollections of this contentious period, the invocation of emergency powers emerged from negotiations between agencies with overlapping interests at the site. Somewhat counter-intuitively, the spill was soon to become a negotiated emergency.

Once defined as belonging to the Coast Guard, it had to be a certain kind of problem, in view the specific goals the Coast Guard was organizationally instructed to pursue and the ways it was supposed to achieve them. However, the spill did not easily fit the Coast Guard's template for spill response. As was touched on earlier, in ambiguous situations organizations have the tendency to invoke solutions that they have used before (Cohen et al. 1972). For the organizations involved with the Guadalupe spill, examining the alternatives entailed taking stock of past oil spills and their responses to them and "favoring" what was already familiar. Aggressive reaction came only after it officially became an "oil spill." By definition, an "oil spill" merits emergency status, as is evident from the following remarks made by a San Luis Obispo Planning Department staffer during a 1996 interview:

When the county started getting involved, in April of 1994, there was more oil on the beach and . . . someone called the Coast Guard, some local person called them and said "Hey there is oil all over the beach, now you guys have got to get involved," and they came up and said "Yes, now we can get involved." They issued a notice of federal interest, then issued a letter to Unocal which directed them to cleanup. "Stop marine releases" is what they said. So it was a whole new ball game. Now it's being treated as an oil spill, because it's in the ocean. . . . As the Coast Guard used to say, when they came in and realized what this was: "Well it's not the usual tanker on the rocks, it's much more difficult to figure out."

The readily available solution required that there be an "emergency marine" release of petroleum into the ocean. The "emergency" designation switched the spill from chronic to crisis status. It allowed the circumvention of permit processes and jurisdictions in the face of an impending "disaster," even if that disaster was the outcome of 38 years of inaction and then negotiation. The new status included an inter-organizational hierarchy that ceded authority primarily to the Coast Guard and the California Department of Fish and Game. The former compelled Unocal to stop diluent from reaching the ocean, leaving the other organizations to stand by and watch. A staff biologist with the California Department of Fish and Game, interviewed in 1996, commented: "In big emergency disasters we use this Incident Command Structure. . . . It is basically just a way to manage huge numbers of personnel and equipment during an emergency situation. . . . It's a way to manage natural disasters, . . . fight fires, do earthquake rescues, and similar things. . . ." In short, the Incident Command Structure, and the powers it wielded under the Oil Pollution Act of 1990, required disastrous implications.

Out of the Coast Guard's actions emerged contradiction and inter-agency acrimony. The initial response was infused with ambiguity. At this point, it was also complicated by rigid and conflicting institutionalized agendas. Instituted to protect marine resources and the environment from acute and disastrous oil spills, aggressive legislation such as the Oil Pollution Act of 1990 promised quick and unequivocal response.[22] However, in view of the kind of problem that Guadalupe presented, regulatory solutions were difficult to arrive at; addressing one set of concerns would mean ignoring another set of equally valid considerations. In this case, ensuring that the ocean was safe from the "imminent threat of petroleum release" meant digging up thousands of tons of sand with enormous earth-moving machines, transporting it across a fragile ecosystem in enormous dump trucks, setting up thermal disorbers (industrial ovens) to treat the contaminated sand, baking the recovered sand at 800–1000°, and transporting it back to the pit from which it had been dug. When rigidly applied, such priorities expose valuations of the environment that have the potential to do as much harm as good.[23]

Bulldozers, front-end loaders, backhoes, tractor scrapers, petroleum booms (to contain already spilled oil), dump trucks, and industrial incinerators are the tools of the trade for the cleanup of contaminated sites. But these are not merely reactionary tools put to use after a problem has been identified. They also reside in Cohen, March, and Olsen's (1972) garbage can of solutions, existing prior to identification and response and shaping what is a problem and how it is addressed. Digging, burning, and replacing tons of sand "naturally" flowed from the spill's emergency designation and the technologies of the trade.

In this instance, regulatory emphasis on the most visible aspects of pollution, notably surface waters and air resources, came at the expense of a more holistic treatment of the environment. The tendency of policy and remedial systems to value certain segments of the environment at the expense of others can ultimately cancel the intended benefit of regulatory

Figure 4.2
Thermal disorbers used in the 1994 beach excavation project.

legislation (Weale 1992; Szasz 1994). Weale (1992, p. 16) notes that "the low priority commonly afforded to some issues of pollution (i.e., soil contamination) and the high priority correspondingly accorded to others (i.e., air and surface waters) reflects the limitations of selective attention built into the processes of policy development." In cases such as the Guadalupe spill, tragic choices were made about what was more (and hence what was less) important in the local environment. These preferences were founded in organizational conceptualizations, not in objective facts about ecology. Just below the surface of these institutionally validated characterizations—revealed through controversy—were organizational approximations of nature: organizational valuations, organizational impressions, and organizational tendencies.

The San Luis Obispo County Planning Department initially criticized the beach dig because it circumvented local permit powers and regulations laid out in California's Environmental Quality Act (owing to its federally determined emergency status). Local officials and community activists protested the beach excavation project because it posed so many unresolved and potential problems. According to critics, because the diluent had been leaching into the ocean bordering the field for some 40 years, an emergency response seemed misapplied.[24] Digging up the beach as if the spill were an emergency looked to these persons like a massive experiment, motivated more by public relations than by a desire to save the dunes.

In essence, the priorities laid out by the Incident Commanders (the Coast Guard, the California Department of Fish and Game, and Unocal) conflicted with an equally important set of priorities expressed by the local constituents. The problems presented by the massive excavation of petroleum-contaminated sand at the beach eclipsed, for many locals, the damages that may have resulted from diluent seeps. Nonetheless, because the Coast Guard felt pressure to do something about the chronic leakage of kerosene-diesel into the surf zone, it devised and rigidly applied, with Unocal and the California Department of Fish and Game, a large-scale industrial project to clean the sand at the beach and "save" the ocean (US Coast Guard 1995b). Waiting and thoroughly assessing a course of action to see if a beach excavation was the best choice was discouraged by organizationally favored solutions. Local environmental groups filed suit to block the proposed dig as a "hastily conceived and environmentally dangerous Incident Action Plan" (US District Court, Northern District of

California 1994, p. 3). In brief, their claim was that the action proposed by Unocal and rubber stamped by federal and state regulators did not match the problem that it was proposed to remedy. This attempt to block the dig failed. Subsequently, the project's Incident Commanders were blasted by local agencies and community members at a public meeting for "acting like a 700-pound gorilla" and implementing a "my way or the highway" approach (lead staff, San Luis Obispo Planning Department, Energy and Natural Resource Division, interviewed in 1996).[25]

There have been other solution-to-problem mismatches in the oil indus-try. For instance, in 1967, when the oil tanker *Torrey Canyon* ran aground off the coast of Scotland, British authorities felt compelled to do something (i.e., anything) as some of the ship's cargo of 35 million gallons of crude oil leaked into regional waters. Their "solution" entailed dropping aviation fuel, incendiary bombs, and napalm on the spilled oil to burn it (Tenner 1998). The "war on the wreck" was as disastrous as the spill itself (Clarke 1990, p. 67; Burger 1997). Similarly, in the cleanup of the *Exxon Valdez* tanker accident, in the quest to get visible crude oil off area beaches, crews used high-power hoses, hot water, and detergents that in retrospect only added to the damages caused by the spill (Tenner 1998; Burger 1997). In both of these cases, "saving" regional coastlines from crude oil entailed throwing the baby out with the bathwater. Though these spills and their cleanup scenarios differed vastly from those approved by the Coast Guard to stem the migration of diluent into the ocean at Guadalupe, their reac-tions reveal a pattern in the selection of solutions.[26]

Similarly, at Guadalupe, the Coast Guard, acting as if it had little time (because once the spill was re-defined as an emergency it did not have *any* time), bypassed due process and directed Unocal to carry out extensive beach excavation. The Coast Guard's powers were based in legislation (i.e., the Oil Pollution Act of 1990) "tailored" to circumstances more appropriate to the *Torrey Canyon* or the *Exxon Valdez* than to the spill at Guadalupe. Whether the Coast Guard's priorities fit the circumstances depends on a hierarchical assessment of the ecological "resources" involved.[27]

Crescive Troubles and "Tankers on the Rocks"

It is necessary to respond to crescive dangers such as the Guadalupe spill as well as to "tankers on the rocks." Many of the environmental problems we currently confront are of a different order than current policies target.

Figure 4.3
Thermally disorbed sand being moved back to the beach in the 1994 beach excavation project.

Incrementally emerging and cumulative hazards are easily overlooked, and they challenge the emphasis of current laws that recognize only acute hazards.

What is more, much like the systemic predilection to respond to acute and immediate impact disasters, the same systems are also prone to favor acute and immediate (impact) solutions. That is, we like our solutions to be as immediate as our disasters. The problem, as should be obvious from the foregoing discussion, is that many of the environmental problems we currently confront will require long-term, if not permanent, management. No solution, as it were, exists to solve them. From Guadalupe to toxic dumps to the storage of nuclear waste, we confront a similar set of predicaments.

Site Characterization and Winter Crises: No Quick Fix

After the beach excavation project of 1994, similar yet less dramatic cycles of crisis and response became an ongoing occurrence for Unocal and the regulatory agencies. While agencies grappled with how to approach "cleaning" the site,[28] they were forced to return annually with the winter storms, applying short-term remedies in hopes of stemming the continued leaching of diluent into bordering river and marine environments. Some have frustratedly felt that they have been "applying Band-Aid solutions" (warden, California Department of Fish and Game, interviewed in 1997) but have done little to permanently remedy the contaminated groundwater or stop it from seeping out. According to the warden just cited tensions continued to play a part in inter-agency interactions, between those that called for immediate and extensive remedial response (namely the Regional Water Control Board and the California Department of Fish and Game) and those that advocated a more cautious approach (California Coastal Commission, San Luis Obispo County Planning Department, San Luis Obispo Supervisors). To manage these winter emergency actions, as well as to address the need for negotiation and coordination, the agencies formed an inter-organizational body: the Multi-Agency Coordinating Committee. According to state and local officials the need for such a coordinating body became apparent in the aftermath of the Coast Guard's 1994 emergency response.

The purpose of the Multi-Agency Coordinating Committee was to move beyond inter-agency conflict and toward a single regulatory voice, or (in the words of a planner with the San Luis Obispo Planning Department, Energy and Natural Resource Division, interviewed in 1996) to "speak with one voice" when managing projects and dealing with an often recalcitrant Unocal. A supervisor at the California Coastal Commission, interviewed in 1997, commented: "[the Multi-Agency Coordinating Committee] was formed because of just how complicated this project really was . . . how we all had to work together on this . . . there was also some concern that the applicant—Unocal—could divide and conquer, by pitting agencies against agencies." However, the difficulties that the agencies confronted at Guadalupe Dunes were more than questions of remedial pace or a resistant Unocal; they included the huge extent of the contamination, the annual winter deluges of water and diluent that were a constant reminder of the dynamic and problematic nature of the site and its cleanup, and, ultimately, the fact that no unqualified short-term solution existed.

The Uncoupling of Problems, Solutions, Participants, and Choices

What has been described in the proceeding analysis is a standard "garbage can" context: The attentions of government regulators were intermittent and unfocused because the problem at hand was ambiguous and did not fit neatly into already proscribed duties. Action was clumsy and uncoordinated because organizational preferences were also unclear. Moreover, participants applied organizational technologies with little reflection on how they addressed the larger issue at hand: the spill and its cleanup. Rather, political expediency—appearing to be doing something—had as much salience as did "solving" the chronic petroleum contamination at the dunes. Finally, because participants and organizations wandered in and out of the decision-making context, accretion of knowledge and continuity were continuously disrupted. All of these contextual factors led to the less than optimal coupling of Guadalupe's garbage can of problems, solutions, participants, and choice opportunities (Cohen, March, and Olsen 1972; March and Olsen 1979). In the case of Guadalupe (similar to Cohen, March, and Olsen's observations of university administrators), because of the ambiguous problem the agencies collectively confronted, routine solutions (for example, emergency response) more readily defined the problem than the reverse (a *solution-to-problem* progression).

More specifically, responding to the Guadalupe spill meant seeing it in a certain light. Early on, the leaks held a to-be-expected status with state and federal regulators. Initially no societal institution existed whose purpose it was to stop, regulate, or enforce chronic land-based spillage. With the 1980s, regulatory bodies were in place to oversee and prohibit threats to environmental health broadly construed (Beamish et al. 1998). Nevertheless, these institutions were created to combat a particular class of *threat*, namely acute forms of industrial crises. Typically, in this context, these are institutionally conceptualized as ocean releases of crude oil—in common regulatory parlance, a "tanker on the rocks." Over time, as signs of the spillage reached a threshold, they became less ordinary and more compelling.

Nevertheless, while new legislation and corresponding institutional actors began to take note of diluent on the beach, they did not aggressively pursue a remedy. At this point the Guadalupe Dunes spill did not meet the criteria for an "emergency" petroleum release, nor was it a benign occurrence. It was perhaps a problem, but not a pressing one. With their predilections (rooted in organizational attributes), regulators were indisposed to see or to act on Guadalupe's chronic leaks until the damage were manifest—indeed, catastrophic.

Once the damage had become threatening, in the winter of 1993, these same institutions began to broach a novel problem for which they had no comprehensive or institutionalized solution on which to base their reactions. The spill crossed jurisdictions and called on agencies to coordinate in ways that had not been done before. For instance, it was not clear just how extensive the diluent contamination was or what kinds of dangers it presented. Moreover, proposing and implementing solutions proved equally difficult and divisive. Solutions involved developing ways of coordinating responses and applying remedies that had to be acceptable to eighteen local, state, and federal agencies and to a community that had become increasingly angry. When it was officially recognized that the dunes were pervasively contaminated and that diluent was leaching into ocean waters, the Coast Guard (with assistance from the California Department of Fish and Game) rushed to initiate the only solution it had for petroleum spills (based on recently passed legislation having to do with response to oil spills: to respond to the chronic wintertime marine releases as if they were an emergency.

Again, the spill's profile was not a high-priority event for regulators and was not amenable to the institutionally formalized remedies (i.e., emergency response). Because of this ambiguity, the Incident Commanders applied the solutions that had worked in the past, but to the "wrong problem."[29] In the end, addressing the spill entailed defining priorities and applying mandates to a case of contamination that no one understood and for which there was no immediate remedy. It was a long-term management project, not the "usual" emergency cleanup; however, emergency cleanup was the only solution that the agencies had in their legislative "tool box."

It became equally apparent that once the spillage was "recognized" the same system was equally inclined to pursue immediate solutions, even though these protocols were a poor match for the problem they confronted. The pressure to solve the field's chronic problems *at once* lent urgency to an already difficult set of tragic choices, or (in the words of a planner in the San Luis Obispo Planning Department's Energy and Natural Resource Division, interviewed in 1996), to a "classic tradeoff": "Do we destroy the dunes to clean up the groundwater . . . or do we manage the groundwater problem?" That tradeoff has continued to confront regulators with each winter's rains.

Postscript

Whether the Guadalupe Dunes oil field can truly be "cleaned" is not certain. What is known of the field's contamination continues to grow worse, while solutions remain elusive (outside of leveling the field and destroying the dunes in the process). For example, sumps containing hundreds of thousands (and perhaps millions) of gallons of crude oil were uncovered in 1997 and 1998. In 1999, PCBs were also found in and around tank farms in the interior of the field (Sneed 1999). For all their surface beauty, the dunes conceal a beastly set of problems (California Coastal Commission 1999, p. 5).

5

A Local Focus: "The Straw That Broke the Camel's Back"

So Guadalupe really . . . was the straw that broke the camel's back. It was just like the community said: "I can't believe they're pulling this again." . . . There was a lot of anger and hatred . . . around that. So in the general sense Guadalupe [represents] "Unocal is screwing us again; they are spilling a bunch of oil into the ground, and they don't want to clean it up. . . ."
—community activist, interviewed in 1997

Local anger toward the spill inflamed anti-oil sentiments in San Luis Obispo County. However, the Guadalupe spill does not stand alone. Appreciating the spill's local status as the last straw requires that one be privy to the socio-historical context within which it, as an issue, is embedded.

As was conveyed in chapter 1, San Luis Obispo County has a distinct history of oil-related activities. In addition, its changing demographics since the 1970s have culminated in a reshuffling of regional priorities. This county, like much of the western United States and especially California, was primarily an agrarian and resource-based economy until World War II (Walker 1998). Over time, it increasingly became a tourist location, a retirement retreat, and, more generally, a place to "escape" urban and industrial growth.[1] Many residents take pride in their county as a place where the hassles of the "big city" can be avoided. Accordingly, an environmental advocate from the county, when interviewed in 1996, attributed his activism to the plight of urban centers such as San Francisco and Los Angeles: "My activism is based on the fact that I had lived . . . in big cities. I came here in 1975 purposefully to get away from those cities. One of the reasons I located here was that it was halfway between Los Angeles and San Francisco, so it can't be *all* polluted." Other informants related similar convictions concerning what San Luis Obispo represented to them, why they came, and why they chose to stay.[2] San Luis Obispans' protective attitude toward their home is deeply rooted in their personal biographies and in their social histories. Connecting these individuals' opinions to San Luis Obispo's history of oil production and its contentious relationship with federal projects makes the community's reaction to the Guadalupe spill understandable.

Defining what "community" means here is important because thus far the spill event has been analyzed from the perspective of organizational involvement. A community is a loosely defined formation of institutions or human associations specific to a location. It is larger than a family, yet less structured and less goal specific than a complex organization. Association is defined by "common belonging" rather than by "pecuniary, instrumental, or utilitarian rationality" (Couch and Kroll-Smith 1990, p. 159).

An Activist County

As in other well-known cases of contamination,[3] community members were the first to recognize and report the Guadalupe spill. In the mid and late 1980s, beach walkers reported suspicious odors and petroleum-stained sands 5 or more years before the whistleblower. A long-time member of the community who frequented the dunes had "noticed pipes . . . with dead [foliage], everything brown and dead, all around them." He had reported the leaks to local Unocal field managers, who had told him that it was "not a problem." Noting such instances and other events that threatened the dunes as a wilderness area,[4] this informant joined with others to form an environmental group called People for the Nipomo Dunes National Seashore. One reason for the formation of the group, he told me in a 1996 interview, was "this little hint all the time of something going on in the Union Oil field—the bad odors, the oil on the surface." In another instance, the director of a local wildlife rehabilitation center called in a slick that had formed just offshore from the oil field. In 1994 a reporter quoted her to the effect that "in 1988 we were telling them there was a problem and nobody listened, nobody did anything about it" (Paddock 1994a). She attributed the death of 65 sea lions and seals to the slick: "They were covered with a shiny coating of petroleum and when you looked at their skin it was just . . . bright red from irritation." (Rice 1994, p. 31)

Notwithstanding these isolated complaints, community activists' initial response to the spill was not one of outspoken objection. Not until the mid 1990s did the spill provoke a concerted effort on the part of activists to protest the dunes' contamination. It would be by way of reinterpretation—that is, connecting the spill to other historical and concurrent regional events—that community advocates would react strongly to what they began to see as an instance of both corporate negligence and governmental complicity.

To lend understandability to local reactions, the following pages relate what Edelstein (1988, 1993) refers to as the *eco-historical* context of the Guadalupe spill. An eco-history, as Edelstein (1993) develops it, directs analytic attention to the derivation of "toxic belief systems" and identities. Edelstein juxtaposes this to communities with no known toxic threats who live in a "normal belief system whereby they do not recognize that exposures can happen at all, or at least not to them personally" (ibid., p. 81). Residents of San Luis Obispo County did not articulate that their environment was a contaminated one; in fact, they generally expressed the opposite: that they would not let their county become contaminated. However, they did share with the communities observed by Edelstein extensive experience with siting controversies, industrial crises, and environmental disasters that have resulted in their being hyper-alert to such issues and exhibiting analogous psychosocial predilections. These include a distrust of industry (the oil industry in particular), an overriding mistrust of federal and state government intentions, and a deep and abiding concern with the potential for environmental mishap.

As evidenced by interviews and conversations with San Luis Obispans, by editorials in the local newspaper, and by environmental advocacy literature, rationally assessing the risks posed by the Guadalupe spill involved more than weighing "probabilistically" whether it would immediately harm individuals. Such a formal rational assessment of the risks involved stands in contrast to the substantive-rational underpinnings of community interpretations of the spill (Kalberg 1980, 1994; Weber 1968). In the case of the Guadalupe spill, "risk rationality" takes on a more historically situated and value-driven tone. Regard for their local environment and experience with the institutions involved with the spill informed these locals of what they could expect. This is an important point of departure as far as risk evaluation is concerned. The very basis of risk assessment as a means of calculating the "objective risks" inherent in a potential hazard is to differentiate "real" from "perceived" danger. Yet this assessment strategy decontextualizes and evaluates risks on the basis of factors too confined to resonate with the community interpretations of "threat" observed in San Luis Obispo.[5] In short, it is the social, historical, and cultural context that provided local residents with a lens through which to see and connect the Guadalupe spill, as an issue, to a larger pattern of institutional behavior. This is in accordance with William Freudenburg's (1992, p. 398) call for "a framework that brings risk to sociology, rather than the converse, highlighting instead of hiding the political and discursive struggles embedded in technological risks." Such a socio-historical context, though not generally a part of conventional assessments of pollution problems, is a part of a holistic accounting of pollution events. Three factors were particularly important in shaping current public responses to the spill: oil's historic place in the county's affairs, the effects of a few precedent-setting events on the region (fertilizing a nascent distrust of industrial enterprise as well as the intentions of the federal and state governments), and the occurrence of several other local oil spills before and concurrent with the (re)discovery of the Guadalupe Dunes spill. After addressing the community's response to these interrelated events, I will elaborate on how San Luis Obispo County's reactions to the spill both extend and contradict what we currently know of public interpretations and responses to industrial crises.

The Environmental Era: An Activist County "Under Siege"

Many who have studied the US environmental movement point to the 1969 Santa Barbara oil spill as one of a handful of episodes that sparked its genesis (Williams 1997; Gottlieb 1993; Kallman and Wheeler 1984; Molotch and Lester 1975; Enloe 1975). San Luis Obispo County's proximity to this massive and dramatic oil spill (just 65 miles north of it) hardened an attitude that was quickly becoming a force in San Luis Obispo County affairs. That the party responsible for the Santa Barbara Channel oil spill was Union Oil (now Unocal Corporation) was not lost on San Luis Obispans. Neither was the name change from Union to Unocal, a permutation rumored to have been related to the negative publicity connected with the older name after the Santa Barbara spill. Santa Barbara's experience reinforced for San Luis Obispans just how catastrophic oil production could be if a community was not vigilant in checking its expansion (Molotch 1970).

On the heels of the Santa Barbara spill came another environmental wake-up call: a proposal by Pacific Gas & Electric, with permission from the federal government, to build the Diablo Canyon nuclear facility in San Luis Obispo County. The idea of a nuclear power plant operating in San Luis Obispo County brought activists and non-activists together in an "Abalone Alliance." Liberals, conservatives, and the "radical fringe" rallied to keep the facility out of San Luis Obispo County in an alliance that was without the usual political or ideological divisions. A series of rallies convened, beginning in August of 1977 and culminating in June of 1979 in a gathering of 40,000 demonstrators, the largest anti-nuclear rally ever held in the United States (Epstein 1991, pp. 96–100). In spite of community resistance, the Diablo Canyon nuclear power plant went on line in 1981.

The lessons the county learned from the Diablo Canyon experience have had long-lasting effects. First, these events fostered skepticism toward the intentions of the federal government and an abiding suspicion of industry. A longtime county activist, interviewed in 1996, conveyed his distrust of industry intentions as follows: "The county experience with Diablo was such that it brought together a vast spectrum from the county, from all walks of life, from all political persuasions, which saw how a corporation will come in and lie and say anything to get what they want. Once they are in, they will continue to lie, breaking all the verbal agreements they have made with the county as things go down the line." Second, many of these activists have since become elected state and county officials, local government bureaucrats, and mainstays in nonprofit environmental organizations. In reference to the effect of the Diablo Canyon siting controversy on sentiments in the county, Nevarez (1996, p. 66) comments: "The protests galvanized a new generation of environmentalists, quality of life-oriented activists, and like-minded officials who remain active in San Luis Obispo government and quite powerful in county politics as well."

While San Luis Obispans struggled with the federal government and with Pacific Gas & Electric to stop Diablo Canyon, on another front the Department of the Interior, under Secretary of the Interior James Watt, proposed a sale of leases for petroleum deposits off the south coast of the county. The already organized activists made an easy switch from one issue of "regional sovereignty" to another—from anti-nuclear to anti-oil protests. The conflation of the nuclear issue with offshore oil is an important one because it parallels how San Luis Obispans would "add up" pollution problems down the road. According to county activists, it was the same battle. They were under siege from a development-oriented federal government that "did not care what San Luis Obispans felt about their coastline" but rather aided and abetted corporate raiders who cared for little but profit (resident of San Luis Obispo County, interviewed in 1997). Similar to the tradition of resistance that John Walton (1992b) observed in citizens of California's Owens River Valley, San Luis Obispans too have developed a tradition of opposition to "urban progress" (ibid., p. 131) as it is articulated by industry and supported by government. In their local struggle, San Luis Obispans have pushed their interests as a local society, with a "culturally rooted sense of injustice" fueling their fight (ibid., p. 131). An editorial in the *Telegram-Tribune* titled "Oil Spill Estimates Buried by Reagan" captures part of the local sentiment and its regard for the federal government's intentions: "New evidence indicates that the Reagan Administration put a lid on some government critics in order to carry out its push for development of oil resources off the California Coast. . . . George Bush plainly is oil company-oriented, but he will have to put his prejudice aside and play it straight if his task force is going to mean anything. He should face up to the possibility that his predecessor was wrong in pushing for an all out oil development off the California Coast. The horrendous Exxon spill in Prince William Sound makes it a brand new ball game."[6]

Together, the 1969 Santa Barbara oil spill, the installation of the Diablo Canyon Nuclear Facility, and the threat of federally backed offshore oil development served to arouse community distrust of federal (and at times state) intentions while marking the oil industry as messy and potentially dangerous. The intensification of local opposition to nuclear power and to oil operations, evident throughout my interviews with local community members, was corroborated in the local press and by the fact that an anti-oil ballot initiative was passed by a majority of San Luis Obispo's voters in 1986 (Beamish et al. 1998; Nevarez et al. 1996; McGinnis 1991).

"Measure A" was a ballot initiative that sought to restrict onshore support facilities for offshore oil development in San Luis Obispo County by making any such development subject to popular approval by vote. Countywide, voters approved the measure by a margin of 53 percent to 47 percent, which on its surface may appear a tight race. By 1990, ballot initiatives limiting the building of onshore petroleum infrastructure had been passed by 26 coastal communities in California; however, San Luis Obispo

County was a special case (as was Santa Barbara County just next door).[7] In contrast with a majority of these other cities, the industry actually produced oil in San Luis Obispo, had a significant presence (regional offices, staff, and local community contacts), organized to stop the initiatives, and far outspent proponents of the measure. For example, in the campaign proceeding the 1986 initiative, the industry outspent local activists in the most expensive campaign in county history. Under the moniker "Citizens for Sensible Ordinances," petroleum-industry interests contributed $426,000 to oppose the initiative that would limit their plans. Voters for "Responsible Oil Development," the anti-oil position, spent $2000 dollars to successfully pass the initiative.[8] The petroleum interests were not done, however. In 1988, Shell Oil proposed to construct an onshore facility. Because Measure A had passed, the project was subject to voters' approval. The industry mobilized in an attempt to gain that permission. Despite industry contributions totaling $275,000 and thousands of handwritten postcards ("Please vote yes on A-B-C. It's the right thing to do.") and only $1000 spent by community activists who mobilized to oppose the project, the proposal was defeated (Beamish et al. 1998, p. 4.2.2).

The local resistance to the development of offshore oil tracts and onshore support facilities continues. However, for the time being, San Luis Obispans have succeeded, through open protest, ballot measures, and support of anti-oil political candidates, in stalling petroleum development off the south coast. A county activist commented on what Measure A represented for him and for local community efforts at keeping oil development out of their county and off their shoreline: "When measure A came through for our county, the county stood up and said, 'Screw you, oil companies. Stay out of here. . . .'"

Other oil mishaps, contemporaneous with the Guadalupe spill's discovery, further spurred and solidified local anger and opposition. They have turned nascent distrust into outright contempt.

The Role of the Media in Connecting Pollution Problems

As described above, the seeds of public outrage about the Guadalupe spill were already present when it made its "public debut" in 1994 via local newspaper and television coverage. Although up until this time the Guadalupe spill had received only minimal coverage, other more "conspicuous" spills received a fair share of the headlines between 1989 and 1993. It was through extensive coverage of those other media events (Fishman 1978) that Guadalupe gained social visibility. As damaging to Unocal's reputation as the spill itself was Unocal's association with an "eco-history" of industrial controversies and environmental accidents. In short, the Guadalupe spill existed not as an isolated event but rather as one element in a larger ecology of pollution events in a county with an existent "political culture of opposition"[9] ready to respond once an environmental problem was identified.

An important part of any public issue's life span and prominence is the media's part in dispensing information about it (Molotch 1979). The media are always an important source of raw data on which future media and private conversations are constructed (Stallings 1990; Fishman 1978; Gans 1980). Moreover, public impressions of risk are intimately tied to media constructions, even if they are not determined by them. Stallings (1990, p. 82) notes that "risk is not the outcome of media and public discourse, but exists in and through processes of discourse."

For their part, regional media outlets, at the very least, have played an important role in keeping oil as an issue on the front pages. For instance, between 1989 and 1996 some 198 stories focusing on oil appeared in the first five pages of the *Telegram-Tribune*, San Luis Obispo's major daily. Theoretically at least, media can generally be counted on to support business and development, especially local media outlets (Molotch 1970; Logan and Molotch 1987). Yet in San Luis Obispo this is not necessarily the case. The *Telegram-Tribune,* and the weekly events and entertainment guide, the *New Times,* are outspokenly in favor of growth control and opposed to oil development. The influence of the *Telegram-Tribune* rests on its ability to set media agendas for regional television stations and for other newspapers, which often follow up on news stories initiated by the *Telegram-Tribune* (Nevarez et al. 1996).

As was related above, outside of direct coverage, an important constituent to regional impressions of the Guadalupe spill have been stories relating and discursively connecting other oil mishaps that occurred just prior to and concurrent with it in the county. This linking of oil spills in the local press was a strong component in the regional outrage that would follow.

The 1989 *Exxon Valdez* tanker accident, much like the Santa Barbara oil spill, alerted San Luis Obispans to the potential for a major spill in their coastal waters. It appeared in local headlines (e.g., figure 5.1), and it was

Boats ride at anchor in front of Union Oil petroleum pier and storage tanks at Port San Luis harbor, where oil tankers often load up.

Oil spill: What if it happened here?

By Mike Stover
Telegram-Tribune

The first fog of summer obliterates all but the distant light at Point San Luis Obispo.

Jigdar picks up an oncoming ship, but for reasons that will eventually land the captain in jail, it goes unnoticed.

The collision comes without warning. The quake and the sound of ripping metal leave no mistake that the impossible is happening.

An oil tanker carrying 10.5 million gallons of crude has crashed, its slippery cargo cascading into the open ocean off Avila Beach.

What happens now? A replay of the Valdez disaster in Alaska's Prince William Sound?

Officials don't draw a pretty picture of what to expect next.

Even under the best of circumstances, only a portion of the oil would ever be cleaned up. And Central California's rugged seas rarely present the best of circumstances.

The area offers a rich diversity of dunes, otters, kelp beds, tidepools, and fisheries.

All would be threatened if a major spill happened, said state Fish and Game biologist Fred Laurent.

This Valdez thing has certainly been a sobering kind of experience.

State Controller Gray Davis agrees.

"The lesson of the Valdez is clear. We cannot rely on oil company representatives that they can handle a spill," he told the Associated Press last week.

More than 200 oil tankers — some loaded with as much as 10.5 million barrels of oil — depart the Unocal and Chevron marine terminals each year. The Unocal operation is located in Avila Beach. Chevron's operation is just north of Morro Bay.

Other larger tankers traveling be-

tween Alaska, Los Angeles and parts beyond pass the county beacon, 10 miles offshore. Each carries the potential for disaster.

That potential will grow in the years ahead if offshore oil development is allowed to move up the coast.

Environmental reports on the six Miguel Project, which will include one platform in the ocean west of Guadalupe, say there's a 1 in 10 chance of a major spill if the Santa Maria Basin is fully developed.

The single platform in the San Miguel

Please see Spill, Back Page

Figure 5.1
Front-page story, *Telegram-Tribune*.

connected to the region by front-page stories such as Stover 1989b. That story goes on to raise the specter of a *Valdez*-type spill in the county, quoting local oil regulators with their appraisals of "how bad it could be if it happened here."

As it turns out, "Oil spill: what if it happened here?" was a somewhat prescient headline. Soon thereafter, Unocal did spill, or more precisely its spills became public events. The first of these spill events occurred in 1992 at Pirates Cove, which abuts Unocal's Avila Beach storage and transport facility. A pipe burst, spilling 6000 gallons of highly visible crude oil into the ocean and onto the adjacent beach. Fulfilling all the criteria that make for drama, the oil spill fouled a favorite local beach and took a heavy toll on a rich wildlife area. The death of 20,000 fingerling salmon ruined a new hatchery program, 62 seabirds died, a dozen sea otters and sea lions were adversely affected, and a lengthy closure of the beach ensued (Greene 1992a).

Following closely on the heels of the Pirates Cove spill, Avila Beach, a small resort community whose livelihood is based on tourism, discovered troubles of its own. After years of denial, Unocal admitted to a "small" problem. The company dug under the city and "discovered" a 400,000-gallon petroleum plume—a mixture of crude oil, diesel fuel, and gasoline floating on top of the groundwater. The spill closed the beach and the town to tourism and devalued local properties. Adding to the plight of property owners and residents who had lived for decades with hydrocarbons under their homes and businesses were fears that their health had been compromised. In 1998, the state of California forced Unocal to purchase most of the parcels of land in the downtown area and to excavate a five-block section of that area, obliterating most of the town.

Playing into the theme of rampant neglect was the resurrection, through local press accounts, of an older Unocal spill at an abandoned storage facility in the city of San Luis Obispo known as the Tank Farm (McMahon 1994a). In 1926, 128 million gallons of crude oil went up in flames when lightning struck some storage tanks. (See figure 5.2.) The spill, according to county activists, would have been avoided if Union Oil had "installed grounding wires" (Bondy 1994). Much of the oil that did not burn still lies beneath Unocal's property. In recent years, the oil-contaminated land has become an issue as the city of San Luis Obispo expands. For instance, the underground pollution has foiled plans by the city to annex the property and build a regional airport; the price tag for cleanup is too high (Wilcox 1994d). The important point is not whether grounding wires would or would not have averted disaster; it is that locals retrospectively ascribe to Unocal a "history of negligence."

Illustrative of the "lumping" effect is a story, titled "Unocal: A Leaky Environmental Record" (Greene 1992b), that recounts all of Unocal's troubles to that date, giving county readers a "rap sheet" of negligence: "It hasn't been a good month for Unocal. For that matter it has not been a good last couple of years. . . . Over the past few years the company has had one

Figure 5.2
Fire at Union Oil Company tank farm, 1926.

environmental problem after another. There's the expensive cleanup of decades worth of gas, diesel, and crude from under downtown Avila. . . . There's the continued friction with neighbors of the company's refinery over odors [, which] have lead to fines . . . and several lawsuits . . . not to mention the underground contamination at its Tank Farm Road facility and another at Buckley Road. . . . The continuing problems . . . are starting to catch the eye of regulators." (See figure 5.3.) A preponderance (76 percent) of petroleum-related articles in the *Tribune* between 1989 and 1996 were spill or pollution stories.[10] Furthermore, if the *Tribune's* coverage of the federal government's locally contentious proposition to develop oil off San Luis Obispo's southern coast and it "oil as pollution" stories are aggregated, the latter represents fully 98 percent of articles that focused on petroleum and related issues oven the 7-year period investigated.

In the other prominent San Luis Obispo newspaper, the weekly called *The New Times*, the inaugural February 1994 story covering the Guadalupe spill (and titled "The Silent Spill") was followed by a story titled "Guadalupe Isn't the First" (Bondy 1994). In a subsequent April issue that recaps the spill, information concerning Guadalupe's pollution problems are sandwiched between and referenced within stories titled "Corporate Crime: Crime Pays Just Fine—If You're Rich" (Fiorenza 1994), "Just When You Thought It Was Safe: Before You Step Foot in the Ocean, You Better

Oil washes ashore at Avila

Tainted wildlife rescued as cleanup effort gears up

By Jan Greene
Telegram-Tribune

The oil slick caused by Monday's Unocal pipeline break covered a quarter-mile square of ocean Tuesday and forced the closure of half of Avila Beach, where tar-like balls began to wash ashore.

Cleanup efforts continued with at least a hundred workers and government officials involved.

Most worked on the bluffs above the spill, just south of the Unocal tank farm above Avila Beach and just north of Pirate's Cove. Others toiled in boats or in shifts on the small, rocky shoreline while the tide was low enough.

About 60 percent of the spill was contained by two long yellow booms, while small boats went after the rest of the oil before it crept out to sea.

Workers also collected four birds that had been coated with the heavy crude. They were taken for treatment to Camp San Luis Obispo, where workers from International Bird Rescue and Pacific Wildlife Care were tending to them. Other small sea animals, such as crabs, were found coated with oil in the tidepools, said state Fish and Game officials.

According to a government scientist at the site, the San Joaquin heavy crude is likely to float on the water rather than settling to the bottom. Over the next few days it will probably begin to turn into tar-like balls and stick to the rocks and beach.

It is also sticking to the thick concentrations of kelp in the area, which may have to be cut away and disposed of. A kelp-cutting boat was scheduled to be on the scene this morning.

U.S. Coast Guard officials overseeing the cleanup said it should take a couple of days to get the crude oil sucked out of the water. They had two ships from Clean Seas, an industry oil spill cleanup cooperative.

Please see Oil, A-8

Wayne Nichols/Telegram-Tribune
Oil cleanup workers ply oil-coated waters near Avila Tuesday as booms corralled about 60 percent of the spill.

Unocal: A leaky environmental record

By Jan Greene
Telegram-Tribune

It hasn't been a good month for Unocal. For that matter, it hasn't been a good last few years.

The company's most recent travails include a sticky oil spill near Avila Beach and possible felony charges following a raid of its Orcutt offices by government officials. They were seeking evidence the

company failed to fully report contamination from its Guadalupe oil field.

But over the past few years the company has had one environmental problem after another.

There's the expensive cleanup of decades worth of gas, diesel and crude from under downtown Avila Beach, which has led Unocal to spend more than $500,000 so far to buy up contaminated property.

And there's the continued friction with neighbors of the company's Nipomo Mesa refinery over odors. Those problems have led to fines from air pollution officials and several lawsuits from workers and people who live nearby.

Not to mention the underground contamination at its Tank Farm Road facility and another at Buckley Road a few years

Please see Unocal, A-8

Figure 5.3
Front-page stories, *Telegram-Tribune*, August 5, 1992.

Read Up On the Link Between Ocean Pollution and Disease" (Decarli 1994), "Unocal's Other Spill: While Guadalupe Dunes Steals the Headlines, Another Huge Underground Spill Goes Unnoticed and Unattended to in San Luis Obispo" (McMahon 1994a), and "Unocal Keeps Popping Up on the Underground Spills List" (McMahon 1994b). The last two stories account for and connect eight local Unocal spills.

Paralleling such media portrayals, informants seldom spoke of the Guadalupe spill without referencing those other controversies or mishaps. "Let me make this point," stated a resident of San Luis Obispo County interviewed in 1997. "There is no difference between the way they are handling Avila and Guadalupe. . . . Unocal's constant lying, foot dragging, misleading, and just their whole game—Unocal has no standing at all!"

Another interviewee, interviewed in 1996, said: "[Unocal] has 75 years of bullying everyone around or buying them off . . . and it didn't happen this time. . . . No one would let them off the hook. . . . It was Guadalupe that unified the community, and Avila came up next. . . ."

The Community and the State: Adding Insult to Perceived Injury

Community resentment of the Guadalupe spill has been exacerbated by an expressed belief that the state and federal governments have, as one county resident characterized it, "dropped the ball." Research on risk and the public trust has focused at length on confidence in authority (Schrader-Frechette 1995; Freudenburg 1993, 1992; Beck 1992a; Giddens 1991; Walsh 1988, 1981). The public trust, according to this research, is easily lost and hard to regain. This is referred to in the literature on risk as the "asymmetry principle" (Slovic 1993). Reconstructing where and when trust is lost, and how this loss effects subsequent events, is instructive, especially in the case of the Guadalupe spill, where locals regularly expressed distrust in Unocal and distrust in the regulatory authorities in the same breath. Case in point: "Initially there was a great deal of frustration locally that . . . not only was the state not doing something, the state appeared to be in collusion with Unocal to put the brakes on the public spotlight. . . . You just don't spill that many gallons of oil and say 'Whoops! Gosh, how about that!'" (resident of San Luis Obispo County, interviewed in 1996)

Other respondents commented on the basic motives, as they understood them, of both Unocal and the government agencies involved. For example: "I think that basically the state agencies and the oil companies. . . . I don't want to say they are sleeping in the same bed, but I think at the time, there was probably not a lot of direction from the leadership on the state level to really look at corporate pollution." (resident of San Luis Obispo County, interviewed in 1997) Once the spill was "discovered," local skepticism toward the oil industry and regulatory agencies grew. The establishment of responsibility for the spill and proposals for cleaning up the diluent provided additional arenas for locals to express their distrust of institutional "outsiders."

Government regulators and Unocal knew that they confronted an embittered community. For example, after emergency excavations at the dunes to stop the migration of petroleum into the ocean in 1994, the Coast Guard initiated an "Incident Specific Preparedness Review" that assessed its spill-response protocols. In this review, which was self-congratulatory throughout, the Coast Guard granted that it needed to do a better job promoting its activities: "In general . . . the state and Coast Guard Unified Command members failed to fully anticipate the intensity of the 'public perceptions' controversy that grew up around the response. Public affairs . . . must be viewed as a critical success factor and must be applied proactively. . . ." (US Coast Guard 1995b, p. vi) "Public perceptions," in this instance, referred to the vociferous opposition mounted by community activists to the Coast Guard's rights under the Oil Pollution Act of 1990 to circumvent local permit approval and allow Unocal to dig up the beach bordering the dunes. Acknowledging the same "community perceptions," a Unocal supervisor (interviewed in 1997) who had managed the cleanup effort remarked on county residents' reaction to the spill and to Unocal's cleanup proposals: "They went nuts! They want to punish Unocal for this spill, so they are willing to take it out on the site. . . ."

Distrust of "outside" organizations, government institutions, and corporations has emerged (like local sentiments concerning oil) from a history with them. In the minds of San Luis Obispo activists and community members, their disregard for the preferences expressed by county citizens have made them suspect.[11] Again, in the case of the Guadalupe spill, the federal and state governments' poor handling of the criminal case against Unocal reinforced this feeling.[12] One community member, interviewed in 1996, expressed "skepticism" that the state and federal governments were looking out for the public interest as follows: "[The California Department of Fish and Game] took the boxes of records and then filed charges one or two days after the statute of limitations. That just boggles my mind. . . . I'm a little skeptical. . . . The statute of limitations had been exceeded by a day or two. . . . I have suspected all these years that [Unocal] paid heavy contributions to candidates that were in their pockets." Such skepticism was a strong undercurrent of the comments of community members who followed the spill, its progression through the courts, and actions taken to clean up the field. At times, community distrust was expressed in characterizations of the state government as favoring big business at the expense of local concerns. In view of the circumstances surrounding the collapse of the criminal case, this point of view is not without foundation. According to community activists, the state's inaction was proof enough of where its priorities lay: "Initially they were not overzealous about cooperating with the county [regulators] and local environmental groups. In other words, . . . the state people would have liked for this thing to not have gotten [any attention]."

The anger directed at federal and state regulators (viewed as "outsiders"), when juxtaposed with the positive remarks about the San Luis Obispo County Planning Department ("insiders'), reveals the impression of irresponsibility, mistrust, and institutional failure expressed by county residents. Consider these two examples:

There is serious politics going on. . . . It took the state a long time, but they have finally just had it with Unocal. . . . The government agencies? The [San Luis Obispo] Energy Department is working their butts off, and they get no respect at all from Unocal. . . . See, all the agencies that are overseeing this are generally conservative . . . except the local Energy Department. . . . At the state level . . . they're just being forced to pay attention to the letter of the (law), not the spirit of the law. . . . Then there is the federal level. . . . It has a mission. It starts with (producing) oil off the coast . . . whose specific assignment is to extract oil. (resident of San Luis Obispo County, interviewed in 1997)

The remedial agencies in San Luis Obispo are sort of activist, not the state, but the county agencies like the Energy Division, are pretty activist, and they are pretty open about it. They are open about opposing oil development and open about insisting that spills be cleaned up. And I think it helps them get things done when there is public outcry. In order for the public outcry to happen, there has to be media attention. And the state . . . that's a different story. (resident of San Luis Obispo County, interviewed in 1996)

The San Luis Obispo County Board of Supervisors created the Energy Division in reaction to the 1981 threat of federally proposed offshore oil development. County resistance to potential offshore oil production was further amplified when it became known that exploitation of those tracts would require an accompanying set of onshore support facilities. The creation of the Energy Division was a means of keeping the reluctant county abreast of petroleum development in their region (planner, San Luis Obispo Planning Department, Energy and Natural Resource Division, interviewed in 1996).

A local beach walker directed his skepticism at what he saw as institutional failure. In his mind, the public was purposefully left out of official legal and remedial processes by state and federal regulators. This left him doubtful that institutional and ameliorative "things are in place to prevent" the Guadalupes of the future: "The anger that we felt early on was at being stonewalled—we thought by everybody involved—by the agencies, as well as the Unocal perpetrator. . . . You still feel anger and you still feel 'Are things in place to prevent that from happening in the future?'" (resident of San Luis Obispo County, interviewed in 1996) These statements reveal a deep-seated distrust of institutional bodies. Researchers and theorists as diverse as Ulrich Beck (1996, 1994, 1992), Anthony Giddens (1994, 1991, 1990), Scott Lash (1994), Brian Wynne (1996, 1992), and William Freudenburg (1993, p. 916) have spoken of such distrust as a defining feature of present-day social relations. Freudenburg (1993, p. 916) refers to this as "the outcome of institutional recreancy." Again, recreancy is the failure of institution invested with the public trust to follow through on their duty(ies).

Michael Edelstein (1993, 1988) has observed analogous forms of recreancy in communities that have experienced siting controversies and instances of contamination. Although Edelstein focuses on hazardous facility siting, some of his conclusions are helpful in understanding the genesis of the institutional distrust that I have observed in the comments of San Luis Obispo County residents concerning the Guadalupe spill. Instead of recreancy, however, Edelstein (1993, p. 76) relates how a community's extensive experience with a proposed or actual hazardous waste facility can result in a stigmatized social psychology that leads to distrust of institutional forms. In his words (1988, p. 195), the hazardous facility siting process "becomes a modern ceremony for selecting victims for sacrifice" in which the common good is impetus and excuse. The psychosocial response to such proposals is a corresponding sense of victimization in both the individual and the collective sense.[13] Those affected see themselves as the victims of "outside," human-based intentionality; thus, they blame the polluter and the government for inadequate prevention and inadequate help. Stigmatization, at a deeper psychological level, also has an associative quality. Being personally or collectively associated with pollution, contamination, or hazardous waste amplifies victim status, further imprinting and disturbing the psyche (Erikson 1994; Edelstein 1993; Kroll-Smith and Couch 1993; Couch and Kroll-Smith 1985).

Distrust, however, is not solely the result of stigmatization. Beyond the potential or actual threat posed by a proposal to install a hazardous waste facility (or by its actual installation), the events surrounding contamination controversies can also engender strong misgivings in a local population

(Freudenburg 1993; Edelstein 1988). Suspicion emerges from feelings of vulnerability: "As residents ponder why the disaster was made or allowed to happen, they question whether government, industry or others had the ability to cause or prevent the exposure, and whether they attempted or intended to do so." (Edelstein 1993, p. 78) Those who confront such threats read their social environment for cues, allocating expectations and responsibility to different actors on the basis of their perceived relationship to the problem at hand.

As can be gathered from the preceding, governments are often targets of this distrust. Governments determine whether proposed activities are permitted. They are responsible for monitoring societal threats and for seeing to it that mishaps are taken care of. Community experiences such as the following are further sources of doubt and community distrust:

• the inability of government institutions to address uncertainty, inasmuch as "experts" often concern themselves with relative risks whereas communities often react to holistic risks (Clarke and Freudenburg 1993, p. 71)

• the inability of governmental institutions to convince residents of the truthfulness of information its claims, in view of previous "failures"

• the inability of government institutions to dispel doubt, due to the pseudo-secret processes that generally shroud decision making in many controversial contexts (Edelstein 1988).

Any one of these experiences is sure to arouse community distrust, yet institutional procedures and responses often entail *all* of them (Brown and Mikkelsen 1990).

The social and historical context of the Guadalupe spill entailed many of the conditions outlined above. From the threat, stigma, and subsequent distrust dredged up during the siting controversy surrounding the Diablo Canyon nuclear facility, to the threats posed by potential offshore oil developments, to the current problems at Guadalupe and associated oil spills, distrust of the intentions of industry, of government, and (to some extent) of outside organizations has become rife.

As we have seen, in the opinions of the San Luis Obispans interviewed for this research the Guadalupe spill is not an isolated event; rather, it represents what I will call an ecology of pollution events. In San Luis Obispo County, the transition from perceiving oil spillage as a decontextualized occurrence to perceiving it as a connected pattern of pollution events mirrors a general trend that has been noted throughout the literature on environment and perceptions of risk and trust.[14] That is, there was a transition from perceiving the spill as a disconnected anomaly—the expected outcome of oil production—to seeing it as representing a pattern of intentional dereliction of responsibility (by Unocal) and duty (by federal and state regulators). In San Luis Obispo, as beach walkers, fishermen, surfers, and others became more aware of issues of environmental "health," interpretations of the "smell" at the beach changed from the notion that it was expected at an oil field ("The oil industry, by its nature, is a dirty industry"—resident interviewed in 1997 and asked about the oil industry in San Luis Obispo) to something that was worth comment or complaint and which provoked distrust (e.g., "I do not feel Unocal was open and above board and straightforward . . . I think they tried to cover it up"—resident interviewed in 1996 and asked about Unocal's response to the spill). This parallels the transition that took place in the 1960s and 1970s from relative security about the environment to insecurity about an environment fraught with partially identified and looming hazards (Szasz 1994; Gottlieb 1993; Weale 1992; Enloe 1975).

Intertwined with the growing distrust of big business and of the government institutions entrusted with protecting the public's well-being, then, has been a commensurate increase in environmental insecurity (Dickens 1996; Lash et al. 1996; Beck et al. 1994; Giddens 1991). In a like manner, the individuals I interviewed about the spill in San Luis Obispo County look at the business practices of corporations and expect the worst. At an empirical level, what was expressed by my interviewees parallels Douglas and Wildavsky's observation that Americans do not seem to trust either the government or corporations (Douglas and Wildavsky 1982, p. 127). Douglas and Wildavsky conclude that this distrust is irrational; however, the concern of my interviewees seems, in view of their collective experience, to be based on sound reasoning.

Is the San Luis Obispans' deep distrust of Unocal and of outside regulators rooted in the presence of petroleum under the Guadalupe Dunes? In the danger to their health and safety?

Although San Luis Obispans have at times disagreed as to how toxic diluent is (or how toxic it eventually will be), questions of toxicity are trumped by strong reactions to what is locally experienced as a breach of trust on the part of Unocal (for having spilled the diluent) and on the part of federal and state regulators (for how they handled the criminal case and the cleanup of the site). The anger expressed by community advocates derived from interpretations of intentionality: that Unocal was responsible for the spill, that the spill was not an accident but rather an outcome of negligence, and that Unocal's local managers denied the spill after it was "discovered." Moreover, adding insult to perceived injury, according to community activists, state and federal regulators were slow to react or to push for resolution—they did not fulfill their obligation to the public trust. These impressions however, did not arise in a vacuum. Community sentiments are anchored to a socio-historical context that bolsters (indeed amplifies) the claims that these institutional others are not to be trusted.

Interpretations of corporate irresponsibility and regulatory neglect have led to a great deal of indignation, inducing reactions that are qualitatively different from reactions to natural and/or acute disasters. First, the pollution at the Guadalupe Dunes is not attributed to natural causes, to the work of God, or to the vagaries of Mother Nature; it is attributed to organizational intentionality and excess. It is seen as a product of Unocal's attempts to turn a profit, Unocal's lack of concern, and Unocal's "save it for a later day mentality" (resident of San Luis Obispo County, 1996). The Guadalupe spill came to be seen as an embodiment of the kind of institutional negligence that threatens the environment and the quality of life more generally. As a symbol of corporate and extra-governmental "dereliction of duty," it has reinforced skepticism toward authority and its aims.

Community, Context, and a "Hermeneutic" of Hazards

My analysis of the Guadalupe Dunes spill shows how a community understood and responded to a major yet deceptively "silent" pollution event. Though with Guadalupe the potential threat to public health is real, the salient feature of the spill lies in what it represents symbolically: betrayal, distrust, malfeasance, and institutional failure. This aligns closely with observations made by Couch and Kroll-Smith (1994) in their research on siting vs. exposure disputes, but for reasons that at first glance would seem to contradict their logic. Couch and Kroll-Smith drew the counter-intuitive conclusion that actual instances of toxic exposure are fundamentally disruptive because the toxin is already present, is disruptive, and is differentially experienced by those that must contend with it. When the hazardous potential of a proposed facility is under consideration in a siting dispute, the source of threat is "outside the physical boundary of the community" (ibid., p. 33). Because siting disputes are outcomes of "beliefs constructed from conversations" about the impact a development will have on a shared domain (the socio-physical space the community shares), they are fundamentally consensus-building events. "It is ironic," Couch and Kroll-Smith note, "that the physical reality of toxic exposure is more apt to create speculation and disagreement than its hypothetical possibility."

Although the Guadalupe spill is physically manifest, the response to it has resembled Couch and Kroll-Smith's articulations of the response to an *exposure dispute* more closely than it has resembled their articulations of a response to *toxic exposure*. First, the spill is far enough "outside" the community as to represent a potential rather than a manifest threat, and its connotations have trumped its exposure threat. Second, while many in San Luis Obispo County have been at odds on a number of important issues (why the spill happened, why it wasn't recognized for such a long time, and how it should be remedied), they have coalesced around the object of their anger. Community groups are unified by their shared outrage, which transcends the spill as a disconnected material event. Instead, county residents have seen the Guadalupe spill as representing the biggest of a number of interconnected destructive events and have judged those responsible on the basis of interpretations of *intent*. This mirrors Couch and Kroll-Smith's assertion (1994, p. 36) that "it is characteristic of siting disputes that a majority of the residents perceive the dangers of the proposed change in strikingly similar terms." This is a hermeneutic conception (i.e., we can only understand the principle of risk in relation to the whole discourse within which it is part), and it is at variance with traditional notions of risk as operationalized in more formally accepted assessment and mitigation methods.[15] In the traditional approaches (still dominant in decision-making circles), the public is left out unless it can articulate grievances in an appropriate manner, generally in terms of the probability of harm or provable sustained damages assessed through oversimplified and isolated notions of risk.

Other important arguments that attempt to explain risk perceptions include some that purport to involve psychological heuristics (Tversky and Kahneman 1974) and some that focus on economic rationality.[16] According to arguments of the first sort, the public miscalculates modern hazards because mental processes are not up to the task of calculating minuscule risks.[17] The public, when "calculating" such risks, tend to employ "computational short-cuts" by overemphasizing events that are particularly memorable (e.g., dramatic events), in effect ignoring events that are less vivid even if they are more common (ibid.). The utility model emphasizes the "[economically] rational, if understandably selfish, response to facilities and technologies that may constitute local undesirable land uses. . . . whatever their objective risks" (Freudenburg 1993, p. 911). As in the other models, a blame-the-victim tone runs throughout. That is, research on risk has tended to "ask what about people leads them to reject certain technological developments, not what about industry leads it to develop technologies people reject" (Freudenburg 1992, p. 399).

Owing to their experiences with industry and government, citizens of San Luis Obispo County do not interpret the risks associated with a hazardous event according to only one method. They use utility, probability, and heuristic psychological devices, depending on the circumstances and the context. Specific to Guadalupe is the fact that, although utilitarian rationality certainly had interpretive play (for instance, economic interests such as tourism and real estate surely did not want the county soiled by pollution), it does little to help understand the whole reaction the spill received. Marginal utility, always hard to argue against in hindsight, is especially hard to argue against when terms such as 'utility', 'rationality', and 'economic gain' remain vague. However, with Guadalupe, economic rationality as a foundation for reactions did not prevail in any of the arguments or diatribes against the spillage. On one hand, the utility of the beach is difficult to put into economic terms such as exchange values. On the other hand, and closer to the mark, environmental advocates from a value-rational position have imputed to Unocal and to government regulators callousness based in their own utilitarian rationality: profit, costs of cleanup, and how much the beach is worth. They are convinced that this is why the spilling went on so long, why it has never received much attention outside the county, and why the community was left out of early decision-making processes. In essence, then, the utilitarianism observed in the goals of institutional forms has helped to lay the basis for the community's interpretations of negligence, not the reverse.

Alone, probabilistic arguments (e.g., those that calculate the probability of death or harm from the spill) are also weak. By any account, the risk of direct harm to humans from Guadalupe was too low to motivate much action. While advocates for the Guadalupe Dunes and community representatives generally felt that the spill was a disaster for local ecosystems, my informants and the advocacy literature tended to dismiss the direct threat Guadalupe posed to their health as the primary reason for their protests. People still used the beach. It was institutional failure across the board that was the cause of their anger. As I have illustrated, this kind of "rationality" was the product of the historical context that informed San Luis Obispans of what they could expect from the institutions and authorities with which they had to contend.

Finally, the psychometric "heuristics" argument does have explanatory power, as is evident from the repeated connections that my informants made between the county's current spill problems and past events. However, in view of its implicit tone, the "heuristics" thesis (i.e., miscalculation as opposed to calculation) does little justice to the rationality employed by San Luis Obispans to understand the Guadalupe spill. Locals rationally identified a pattern of misconduct, and they behaved accordingly. Owing to their roots, these theories of risk perception would have San Luis Obispans address nuclear power, offshore petroleum development, and petroleum spills as isolated issues. San Luis Obispans have not gone along with that. Instead, in an interpretive strategy that none of these formal risk analyses admit as a valid or even a rational concern, they have connected the Guadalupe spill to their larger struggle to ensure the health and well-being of their home.

Much as in the other social settings related in previous chapters, response to the spill has derived from elements that have more salience to locals than do the millions of gallons of petroleum that have accumulated underground in an isolated corner of San Luis Obispo County. The conspicuous features of this spill have included the historical context within which oil production occurred, cultural norms that guided behaviors, and institutional mechanizations that bounded individual and organizational rationality in important ways. Specific to the community's response, the epicenter of anger was not confined to the singularity represented by one of the nation's largest recorded petroleum spills; it also involved a socio-cultural context that included history, community identity, and a sense of betrayal.

6

Staring Blindly at a Problem: Incrementalism and Accommodation

This is the way the world ends
This is the way the world ends
This is the way the world ends
Not with a bang but a whimper.
—T. S. Eliot, *The Waste Land*

Now that the specifics surrounding the making, the "discovery," and the community's response to the Guadalupe spill are analytically understood, a theoretical synthesis of these findings is in order. In my extensive analysis, I have used many concepts outlined in research on human-induced environmental hazards, organization studies, and theories of social change. However, it is clear that a model is needed to address circumstances not yet articulated in previous investigations. The problem(s) the Guadalupe spill exhibited diverged considerably from those explored by earlier authors, both in the kind of physical aberration the spill represented and in the reactions it evoked. It was chronic, crescive, and unspectacular. It was based in simple technologies, it involved a loosely coupled organizational setting, and it has no comprehensive solution. Moreover, it inspired little dread but a great deal of local outrage. In short, the response the spill drew from all those involved was reminiscent of the parable that opened my introductory chapter: like frogs, human systems accommodate chronic, crescive, and diffuse problems until they become disastrous.

In the early stages of my research, it seemed that each social setting had its own unique response to the spill. However, deeper examination revealed a pattern of non-response and organizational maladjustment based in sociocultural and organizational normative drift. Understanding 40 years of tolerance entailed addressing how inertial forces produced by social organization had desensitized those involved to accumulating signs of crisis (Beamish 2000). As I outlined in chapters 3–5, social relations bind human actors to patterned ways of knowing and interpreting events. In the following pages, I will explore this social dynamic and how it propagated tolerance for the sight and smell of petroleum at the dunes.

Social Order and Social Habituation

Social organization in both (overlapping) cultural and institutional forms involves inertia that (re)produces the foundations for social stability. As social science has used it, inertia as observed in cognition, interpersonal interaction, and complex organization is a product of consistent and repeated acts, routine conduct, schemata, traditions, conventions, or habits of practice (Camic 1986; Bourdieu 1977, 1984). As Camic (1986) points out, the concept of habit, which has a long history in the discipline of sociology, was long ago shelved as a conceptual and explanatory variable to address social stability. However, some attention to its conceptual roots is important to a thorough understanding of its meaning as it is evoked in the following pages, insofar as the concepts of habit, inertia, and stability developed in this study are not to be confused with the concept of habit as it is operationalized in behavioral psychology, in physiologically, and in biology. In the usage of those fields, 'habit' denotes a motor reflex, entirely outside the social world, that is rooted in the excitation of nerve cells and observed in physiological and autonomic responses.

   The social theorists Durkheim and Weber, while in disagreement on most of what one would call theoretical substance, used habit and routine extensively in their early works. Yet both Durkheim and Weber—and with them sociology—expeditiously dropped these concepts as they became the intellectual property of behavioral psychology.[1] More recently, Camic (1986) and Bourdieu (1977) have re-fortified the concept of habit by connecting it with institutionalized norms and the resultant social structural inertia(s).

   Bourdieu (1977, p. 74) develops a "theory . . . of the generation of practices" that elaborates on the inertial qualities found in social reproduction. Bourdieu reintroduced the concept *habitus* to explain the self-generative disposition of habit and convention, through which social history and social class appear as human nature.[2] According to Bourdieu (ibid., p. 79), tacit cultural assumptions function at the subconscious, emerging as structure but metaphorically representing a discourse that "feeds off of itself like a train bringing along its own tracks." The train is culture, steaming ahead into the future; the tracks are cultural habits, tacit assumptions, and the collective unconscious, which reproduce social distinctions and predict social actions.

In another theoretical work on the habitus-inertia linkage, Becker (1995, p. 301) observes that "one of the remarkable things about the world of classical music is how stable it is through time." He explains this consistency by describing music making as bound up in conventions that result in hegemonic practices. It is through the minutiae of music making (the traditional shape of an instrument, the notes it can emit, how one learns to play, the phraseology employed by musicians, and so forth) that social inertia manifests, carrying convention into the next moment and thereby circumscribing performance in the future. To put this another way: Social relations, in their totality, are inscribed in both material technologies and cultural routines. This imbues them with consistency over time, and it is evidenced in customs and tacit assumptions that are resistant to modification for immanently social reasons: the potential for coordinated human interaction pivots on both the stability and predictability of the systems within which humans reside. This is similar to the way sociologists of science and technology describe "sets of ideas," "packages," or what Latour (1987) describes as "lash-ups." In Latour's language, an accepted object such as a scientific fact or technology requires a complex assemblage of active allies, procedures, and hardware that, once woven together, sustain it through time.[3] Each aspect of the package, be it in music or in a scientific enterprise "presupposes the existence of all the other pieces" (Becker 1995, p. 304). Again, the "social package" exerts inertial force. The power exerted by already manifest relations between parts raises the price of deviation (i.e., innovation) from the routines, protocols, and taken-for-granted assumptions that constitute the apparatus of a corporation, a regulatory agency, or something as seemingly indistinct as a "community."

The intent here is not to pathologize the inertia inherent in human systems but to problematize it. On the one hand, inertia of this sort is necessary, giving human relations the predictability and stability needed to make interaction possible through time. On the other hand, it reveals a proclivity to become inured by habit.

Organizational theorists too, with somewhat differing interests, have described inertial qualities represented in corporate forms, culture, and routine practices. While maintaining a strict organizational focus, the categories used by these theorists parallel those in other realms of social interaction.

According to the organizational ecologists Michael Hannan and John Freeman, complex organizations (or, more generally, corporate entities) require social and material investment to sustain them. Organizational builders must "accumulate capital, commitment of potential members, entrepreneurial skills, and legitimacy" (Hannan and Freeman 1984, p. 152; Stichcombe 1965). However, these sunk costs are not a one-time proposition. A substantial portion of organizational resources must be constantly reinvested to ensure continuous corporate integrity: "Just as in the case of biotic creatures, there is a substantial metabolic overhead relative to the amount of work performed." (Hannan and Freeman 1984, p. 153) The cost of creating a permanent organization is steep when compared to other alternatives, such as free association or ad hoc collectives (ibid., p. 154). Thus, the following question arises: "Why do individuals and other social actors agree to commit scarce resources to such an expensive solution to problems of collective action?" (ibid., p. 152) The answer Hannan and Freeman put forth, much like the others, hinges on the predictive value (e.g., behavioral predictability) of reliability and accountability. Hannan and Freeman discuss forces loaded against change and innovation.[4] These forces, they argue, are fundamental to inter-organizational and organizational-societal exchanges. The first, reliability, lends normative overtones to organizational endeavor. "In a world of uncertainty," Hannan and Freeman note, "potential clients may value reliability of performance more than efficiency" (ibid., p. 153). Accountability, on the other hand, addresses the requirement that an organization be able to rationally account for its actions. This does not mean that an "organization must tell the truth to its members and to the public about how resources were used or how some debacle came about" (ibid., p. 153), but it does require that the organization's argument remain internally consistent. Moreover, an organization must continually reiterate these qualities with "high fidelity" in order to maintain intra- and extra-organizational legitimacy. The tradeoff for such predictability, according to Hannan and Freeman, is an increase in "structural inertia": "If selection [organizational evolution] favors reliable, accountable organizations, it also favors organizations with high levels of inertia. In this sense, inertia can be considered to be a by-product of selection. Our argument on this point may be considered an instance of the more general evolutionary argument that selection tends to favor *stable* systems." (ibid., p. 162)

For the purposes of this discussion, it is the durability and the outcomes of such processes—cultural and organizational practices that produce accommodation—that hold explanatory promise. Though the theorists cited above may disagree on the amount of agency, structure, and socialization that instigates social action, and on the benefits and/or liabilities of each, they all attribute extraordinary influence to convention, routine, social investment, and re-enactment and to the effects of inertia on human systems and the psyche. In short, social inertia provides a way of understanding "ignorance" without dismissing it as simply pathological.

The Guadalupe Spill as Habit

Although they are common excuses for events such as the Guadalupe spill, avarice, deception, and dereliction of duty are too simplistic to explain away 38 years of grievous leaks that many individuals and organizations, with divergent interests, knew about during that time. Remember, by the 1980s these leaks sometimes amounted to more than 8000 gallons per day. Moreover, signs of petroleum were noticeable to people who frequented the area for recreational purposes for at least 10 years.

For conceptual clarity, it is important to break the response to the spill into discrete periods, bearing in mind that some characteristics of particular moments were at play throughout the duration of the spill.

From the 1950s to the Early 1980s

During the first 30 years of spillage, accommodation to the chronic releases of petroleum reigned across social settings. Oil workers at the field saw small leaks as a normal part of production. To pin the lack of early recognition and the later lack of confession entirely on the profit motive or on the negligence of individual operators would obfuscate the confluence of factors that initiated the spill and its eventual coverup. As was outlined in chapter 3, the answer is located in a set of unremarkable organizational attributes that characterized oil production and bounded the rationality of workers at the field.

Early on, the ignorance surrounding the ongoing leaks and spills demonstrated the power of "lashing together" repetition, convention, and context in creating and fortifying strong normative and cognitive frameworks. These schemata blunted perceptions of ongoing and destructive events. This also lends insight into why oil workers who had initially accepted the leaks and spills as part of the job would eventually resort to silence as a means of avoiding the problem under their feet when they recognized its magnitude.

Authorities also acted as if the spillage was normal. Initially, no societal institution existed to stop, regulate, or enforce this kind of land-based spill. By the 1980s, regulatory bodies were in place to oversee and prohibit threats to environmental health broadly construed. Yet these institutions were created to combat earthquakes, hurricanes, floods, "tankers on the rocks," and other immediate disasters. Their solutions, in the main, are post-disaster reaction and reconstruction. Agencies' protocols and response regimens are not disposed to recognize crescive troubles until damage is manifest. The Oil Pollution Act of 1990, though it encouraged response to and remediation of large ocean-based oil spills, also had the unintended consequence of promoting passivity on the part of regulatory authorities by bounding their fields of attention. At first, government regulators failed to acknowledge the Guadalupe spill because it was not an emergency, and they had no ready-made response for such a chronic problem. In the initial court depositions was to play up the fact that this left Unocal to self-report. Grilling investigators, Unocal's lawyers insinuated that blame for the spill lay with regulatory inactivity and implied that if regulators had been doing their jobs the spill would have been remedied long ago (US District Court 1994). Yet, unless an event poses obvious crisis potential, there are few regulators in place who have the authority, the resources, or the time to actively investigate. In short, regulatory agencies focus on a limited portion of an often-complicated sequence of events. Whether an event can be defined as an emergency or whether it has no immediate consequences plays a significant role in whether (and how) it will be recognized and acted upon.

Many residents of the community—at least those who walked, surfed, and fished the beach—knew that the area smelled of petroleum. Surfers thought it was normal that their eyes burned at times, that the water had a sheen on it, and that the beach smelled. The Los Angeles Times (Connell 1999) quoted one local resident as follows: "Years ago, we used to dig for worms out there to use for bait and we'd find all this oil in the ground. The smell was something, but I didn't know. . . . I sure don't fish there anymore." Over time, the normalcy of spilled petroleum, acquired cultural and institutional inertia.

The Late 1980s

With changing definitions of environmental health, and with the introduction of government agencies to enforce these new definitions, the spilled diluent that had once been unproblematic became a source of cognitive dissonance. The whistleblower became uncomfortable with the spillage when he began to acknowledge it as a problem. The regulators who responded to reports of oil on the beach acknowledged the trouble only after its acuteness became clear.[5] A Regional Water Quality Control Board inspector I interviewed in 1996 related that by the winter of 1989–90 the leaks no longer appeared to him to be an "ordinary spill." The community, which was undergoing the same changes in environmental sentiments, also began to realize something was not right at the Guadalupe oil field and the adjacent beach. Community members began calling the authorities to report that the beach was smelly. The period of accommodation was over. The problem was no longer normal by the prevailing definitions.

Into the 1990s

Once the spill began to trouble some who frequented the beach and the dunes, normative ways of knowing worked to increase the price of innovation (cognitive, social, and complex organizational), even in light of the distressing circumstances they observed. In tracing knowledge of the spill within and between the settings, I observed the ensuing struggle over what to make of it—a struggle that was based in the personal and organizational relations of the parties. At the field, Unocal's local reaction to the spill (both what had already spilled into the dunes and what continued to spill until 1990) was marked by social and organizational inertia located in the field's work structure, in the culture surrounding production, and in the oil field's isolation from outside interference. Over decades, production routines had formalized the normality of daily spillage, including how interactions with regulators were conducted.

Organizationally speaking, once the spillage became problematic, reporting it was not the responsibility of individual field workers. Superiors were responsible for reporting such things. Furthermore, if a worker were to break rank and report the spillage, he would risk being fired, not being promoted, or generally being blackballed by co-workers and by the industry as a whole. Thus, when the accumulation of diluent became troubling (even to those at the oil field), denial as a "solution" emerged from the field's organizational culture.

Socially, and to a degree economically, it was the workers' field. If they admitted the spill, and if the field was shut down, they would lose their jobs (as eventually happened in 1994). Cultural solidarity also greatly contributed to in-group cohesion and further silenced workers, lowering the chances of an out-group admission. In this manner, the secret of the big spill under their feet was kept inside for everyone's (individual and collective) interest. Once the spillage had been problematized, normalization turned to denial and denial to silence for the sake of individual and organizational preservation.

The "dark side" of a strong normative culture was evident at Guadalupe. None of the individuals who knew about the field's pervasive (albeit undramatic) petroleum contamination responded until it was an established and incontrovertible disaster. When the danger the spillage presented was recognized, the workers did not turn to remedy, but rather to secrecy and coverup. The ability of workers to keep information inside field operations is attributable to the field's social organization and cultural milieu and its structural isolation. These factors (along with Unocal's corporate "disposition," which did not stress environmental compliance) explain the long-term continuation of the leakage even after workers recognized the leaks and spills as a significant problem.

For the regulators, because it wasn't "a tanker on the rocks," the spill was not amenable to the solutions that were (and still are) institutionally formalized. Thus, when confronted with the accumulation of diluent, the involved agencies rushed in and initiated the only solution they had on the books: to respond as if it were an emergency. Innovation was out of the question. In other words, the lead agencies re-framed the spill so as to be able to respond in a way that made it sensible to them, applying solutions that had worked in the past to the "wrong problem" (chronic trouble).

The community had no dramatic spill around which to mobilize. Aside from its volume, the spill (initially) provided little incentive for outrage—it did not disrupt daily life in a way that inspired organized social protest.

It was underground. Its effects on the site and the surrounding region were ambiguous. On its surface, the beach looked wild and healthy, even if it smelled of diluent at times and even if the river and the ocean showed signs of petroleum stains. The spill was seen as a problem only when environmental advocates, the local media, and community residents discursively connected it to other local Unocal spills[6]—that is, when the community saw it in a larger eco-historical context (Edelstein 1988).

The local interpretations of institutional negligence is important to understanding the community's response to the Guadalupe spill, which was based in culturally rooted skepticism toward "outsider" interests.[7] The community's anger was the outgrowth of past experiences with federal and state agencies that had promoted distrust. The community saw the Guadalupe event not as a single spill but as one of many in an ecology of pollution events that represented both corporate negligence and "a failure of state and federal governments to carry out their responsibilities with the degree of vigor necessary to merit social trust" (Freudenburg 1993).

Inasmuch as community reactions articulate a latent distrust of institutional forms, they parallel qualities expressed in theories of modernity as developed by Beck in *Risk Society* (1992) and by Giddens in *The Consequences of Modernity* (1990) and *Modernity and Self-Identity* (1991). Beck and Giddens converge on the idea that the defining feature of late modern society is an abiding distrust of institutional forms and on the idea that "providential reason" is no longer taken for granted (Giddens 1991, p. 28). The idea that "increased secular understanding" and hence control of nature "intrinsically leads to a safer and more rewarding existence for human beings" (ibid.) has been superseded by a societal outlook in which the "unknown and unintended consequences" of institutionally sanctioned techno-scientific and industrial endeavors "become the dominant force in history and society" (Beck 1992a, p. 22). Beck (1996, p. 56) expounds on "risk society's" break with the modern project, in which trust in progress (and in its institutional advocates) is no longer taken for granted: "Modernity's quintessential institutions of technology and science self-refute their own enlightenment promises and programs. In particular, the failure of institutions to control the risks they have created, seen most accurately in the ecological crises of industrial society, has generated a more profound and pervasive sense of risk. As the contradictions grow more frequent and intense, so the sense of risk grows, and the legitimacy of those institutions which have designated themselves as saviors erodes correspondingly."

Both Beck and Giddens, but especially Beck, put forward the idea that in high modernity the fear of institutional malpractice holds more salience than more "natural" forms of environmental threat. In short, contemporary society is simultaneously trusting in and suspicious of expert systems.

Yet in the story told in these pages there is something more than either Beck or Giddens recognized in their theoretical deliberations on risk society or high modernity. Out of an undramatic case of creeping contamination, a great deal of mistrust was generated that in many ways is as ancient as it is modern. This spill was not a Bhopal or a Chernobyl, "for which no insurance is available" and from which, according to Beck (1996, p. 31), modern anxiety, mistrust, and risk have emerged in risk theory. The spill and the non-response it gained were based in age-old organizational, cultural, and ideological elements coupled with very human shortcomings: myopia, sloppiness, and (at least in the 1980s and the 1990s) outright criminal negligence. The distrust that San Luis Obispans articulated, then, was not the outcome of a disaster that was going to ruin their lives, at least not directly. Rather, the sense of imperilment (i.e., risk) stemmed from their mistrust of the responsible institutions, which was based in a local sense of betrayal. Furthermore, as expressed by San Luis Obispans, this engendered uncertainty, suspicion, mistrust, and a good deal of local anger. Ultimately, this sense of betrayal is a basis for reassessing the future and hence for the public's recalculations of risk (Freudenburg 1992, 1993). Put another way: Institutions that held positions of "trust, agency, responsibility, or fiduciary or other forms of broadly expected obligations to the collectivity" (Freudenburg 1988) failed to follow through on their "duties" in ways that eroded public trust in them.

Trust is a very old constituent of community building, and breaches of it provide ammunition for social movements. According to the labor-movement theorist Craig Reinarman (1987, p. 18), the postwar social charter is "nowhere written" but rather represents "tacit rights and expectations" that have been "central to the legitimacy of the US political economic system throughout the [postwar] era."[8] The violation of this social charter between the general citizenry and the responsible organizations arose from the federal government's and Unocal's repeated reneging on a deal they had struck with society. If Unocal were truly "Committed to the Community," as it has repeatedly claimed in promotional campaigns to remedy its locally infamous corporate reputation (Unocal Corporation 1996a,b, 1994b, 1993), it would not have behaved as it had over the years.

The lack of a response from the government trustee agencies that were responsible for ensuring the public safety appeared to the community as incompetence or even collusion. That the agencies failed to pursue the spill expeditiously even though they had been aware of it for some time was not lost on local citizens, who repeatedly told me that they had lost faith in these institutions to follow through and protect their local interests. The San Luis Obispans' reactions were based more on institutional negligence than on direct signs of the spill or on the threat it presents to their health.

Crescive Trouble and Collective Response

Although interpretations of impact and solutions may differ for each social and institutional setting, the response sequence has been remarkably similar across them (table 6.1).

During the first 30 years of the Guadalupe spill, social inertia in organizational and individual habituation and acclimatization to incremental changes (repeatedly spilled and accumulating petroleum) that left field personnel, regulators, and community blind to the spill as an environmental threat. As time passed, normative boundaries expanded (stage 2 in the table) to include greater amounts of petroleum as routine. Once a critical threshold had been met in the 1980s (stage 3), what was until that time normal became troubling. A period of denial ensued (stage 4) followed, insofar as there was no institutionalized way to address 6 square miles of

Table 6.1
Stages of response to Guadalupe Dunes spill.

Stage	Characteristics	Period
1	Accommodation, acclimatization	1950s–1970s
2	Expansion of normative boundaries	1970s–early 1980s
3	Critical threshold	late 1980s
4	Denial	1990–1992
5	Conflict, reframing of event into sensible terms	1993–1995

petroleum-saturated sand dunes. Finally, those involved sought to re-frame the event (stage 5) in psychosocially manageable terms.

The spill is technically insoluble because its effects on the region, in both the short and the long term, are still matters of extensive debate, remaining largely unknown (one cannot fix what one cannot characterize) but unquestionably negative. The spill represents an environmental double bind: Not only is it prohibitively expensive to fix, but cleaning it up may be as destructive as leaving it in place. This has left much of the process open to rivalry among competing frames of reference, each advocating for solutions couched in conflicting scientific claims that are ultimately based in values, vested interests, and wistful hopes.

Negotiating Ecology through Policy

The preceding discussion and the research on which it is based reveal a significant predicament facing American society. Geared for acute crises, we are inclined to stand by as circumstances deteriorate. Where once popular conceptions of "the end" (especially during the Cold War) posited a dramatic and fiery finish, cumulative, interconnected, chronic, and synergistic degradations now provide the doomsday scenario. Case in point: Beginning in the late 1960s, the United States saw a flurry of environmental activity that culminated in environmental policy and law. Since 1970, this institutional environmentalism has been primarily pollution-based and law-driven (Chertow et al. 1997). Although a great deal of headway has been made in mitigating first-order environmental threats (natural and industrial), much remains to be done in identifying and remedying more diffuse and less obvious forms of environmental degradation. Current statutes have been fairly successful in raising standards for industrial producers and cleaning up water sources (e.g., lakes and rivers) and air pollution districts (e.g., the Los Angeles Basin). Nevertheless, as the Guadalupe spill attests, and in view of the problems we now collectively confront, continuing down the path laid out by these policies seems inadequate.

Many of today's environmental problems are of a different order than the problems that current policies target. Incrementally emerging and cumulative hazards analogous to the Guadalupe spill[9] are more difficult to pin down than the acute crises that motivated many of the current laws. These crescive troubles—and troubles far beyond the confines of "environmental problems"—cannot be left to self-reporting, cannot be addressed through traditional and atomized conceptions of the problems at hand, and cannot be solved through a conventional "react and rebuild" mentality.

For instance, the limitations of current legal and institutional regulatory systems are apparent in their compartmentalized approach to problems. Human communities, as well as air, water, land, flora, and fauna, are treated as separate components under a complex structure of laws that are rigidly applied to problems as they arise. This partitioning is mirrored in the boundaries of the academic disciplines through which questions are asked and answers forwarded. The results are legal and institutional gridlock and environmental degeneration (through unremedied discharges of effluents and/or through the eventual and often impact-loaded cleanup of such discharges). Stemming and managing environmental crises entails more attention to factors that predict their genesis than is now being paid, as well as a more cooperative, less fragmented, and more comprehensive approach to their mitigation.

There is no doubt that this will not be easy. It calls on us to develop a "systems" approach that combines rigorous analysis with an interdisciplinary bent and that accounts for social context as a matter of policy. A "systems" approach entails widening our view of what matters from an almost exclusive focus on "what comes out of the end of the exhaust pipe" to a more proactive and integrated focus that addresses signs of impending crises as they incubate. More often than is currently the case, environmental analysis should involve accounting for how identified hazards come to be recognized—how threats were taken note of, problematized, and acted upon (or not).

Pollution, according to the policy analyst Albert Weale (1992, p. 3), is "the introduction into the environment of substances or emissions that either damage, or carry the risk of damaging, human health or well-being, the built environment or the natural environment." Perpetual dissension over what is and what is not damaging makes it more important to understand how pollution, as a social and physical event, is named and responded to. As the story of the Guadalupe spill attests, it would be misleading to seek a purely "objective" point from which to make decisions concerning the spill or its cleanup. Environmentally based hazards must be perceptually processed and problematized before they are seen and acted on as such. People do so in situ. In the case of the Guadalupe spill, to describe this naming process is to delve into the nexus of historical, cultural, and organizational contexts, as well as the biophysical attributes the spill exhibited. This makes it possible to understand why the spill went unnoticed, unreported, and unmitigated for so long. Through such an understanding, more proactive systems of response could be developed that would take heed of actors who, through incremental change and normalization, no longer perceive their environment as dangerous and worth comment—whether they are workers, regulators, or community members.

The real consequences that we and the environments we inhabit face from instances such as the Guadalupe spill should not be seen as relative even if they are, in part, socially constructed. Such occurrences and their legacies are a universal and destructive aspect of post-industrial landscapes. The hard reality is that environmentally destructive trends continue unabated. Yet human systems selectively choose what to attend to and (for lack of a response) what to ignore. Increasing our understanding of the interpretative processes that surround ignorance and acknowledgment of threats is a necessary step toward remediation.

Appendix A
Methodological Remarks

A Narrative Case Format

According to Walton (1992a, p. 121), "when researchers speak of a 'case' rather than a circumstance, instance, or event, they invest the study of a particular social setting with some sense of generality," and the relevance of a case lies in that it is representative of a family or a genre and as such it makes implicit claims to a "general category of the social world." It is in this sense that I explore the Guadalupe spill—as a general category, a class of ecological crises based in and through social relations—while always keeping an eye out for more general and far-reaching implications. While addressing how an environmental crisis was generated and responded to, I also address issues and concepts that relate to problems and to social change more generally, ranging outside the confines of an "environmental dilemma."

Induction and Exploratory Research

I began participating in research on the Central California petroleum industry in 1990 (Beamish et al. 1998; Molotch et al. 1998; Molotch and Freudenburg 1996). Specifically, I explored the creation of and response to the Guadalupe spill over a 2½-year period, commencing in 1996. The questions I asked of the case and the issues that emerged also called for historiographic attention. As I probed the spill's "discovery," I found that local impressions of it had their underpinnings in socio-historical relationships and contexts such as the economic place of oil production in the region; the changing demographic profile of the county; the evolving federal, state, and local legislative and regulatory agendas; and the transformation of the petroleum industry. Thus, as I collected and analyzed data, it became paramount that I address the spill, as an issue, with these historical elements in mind. This history provided the current state of affairs with continuity; it imbued the spill, via its local connection to other issues and events, with meaning that was "larger" than if it were an isolated occurrence. In view of the questions I was asking, it also meant that multiple data sourcing was a necessity, for no single source would give me the kind of breadth and perspective required to reconstruct the social and historical elements involved.

The data collection and analysis I went through mirrored closely the logic that Gould (1986) describes as underlying a historical science and its reliance on the consilience of induction. Consilience literally means the accordance of two or more inductions drawn from different groups of phenomena over time. Though the method itself (as Gould articulates) emerges from evolutionary biology and paleontology,[1] the tenets Gould explicates are useful when historically reconstructing events that are the result of overdetermined social processes. This is not to be confused with triangulation, which relies on a similar set of premises, but holds a different methodological emphasis. In my case, using analytic induction, I pieced the story of the spill together (again as an issue) by relying on the coalescence of multiple sources, as opposed to verifying separate sources (or angles) and then pulling them together to make a stronger case (as with triangulation). Although the two methods contain conceptual overlap, the difference is in the one (consilience) stressing "proof" through the accumulation of small-scale empirical observation based in a diverse array of data, as differentiated from the other (triangulation) that calls for the use of two or more means of assessment to more accurately measure a phenomena. The emphasis Gould makes (and that I applied in the course of this research), when he describes Darwin's historically based science, is one oriented toward observing gradual, small-scale changes as the bedrock on which the immense phenomena of history are built.

It has been in a like manner—through the identification of iterated patterns, across a wide array of data—that I have drawn conclusions. Thus, to reconstruct the event, I conducted field interviews with respondents drawn from a range of groups: community members, environmental activists, government regulators, and members of the local oil industry. I also pursued related ethnographic contexts, amassed and analyzed a substantial assemblage of archival materials, and closely followed media portrayals beginning in 1989. Taken separately, none of these sources of data could provide conclusive substantiation; taken together they represent a preponderance of evidence that revealed homologies likely missed otherwise. Below, I describe in some detail each of the data sources I explored.

Interviews

In the first instance, I conducted 39 field and telephone interviews with informants drawn from four primary groups:

• community members who, while not necessarily directly linked to activist groups, had specific reasons for interest in the spill—such as beachgoers, media employees, local historians, and others who, by virtue of physical proximity, recreational interests, or special roles as chroniclers of local events, seemed likely to have insight into how the spill was "felt"
• government regulators (including those directly involved in cleanup, litigation, and policy construction) and relevant elected officeholders
• environmental activists who had actively pursued an agenda that countered the interests of the local petroleum industry
• members of the local oil industry.

I had access, through a colleague also studying the region's oil industry, to three other transcripts in which workers from San Luis Obispo commented on Unocal, regional oil production, and related matters. All these groups provided a "ground-level" understanding of how the spill was received. Thirty-five of these interviews were tape recorded, transcribed, and systematically analyzed. Four informants were uncomfortable being taped, so interview notes and my memory represent the interactions. The quotations cited throughout the text are verbatim from the transcripts or from interview notes unless cited otherwise.

Interviews were first held with people whose names surfaced in the media or in official documents and who were representative of the groups of interest outlined above. After interviewing a respondent, I asked for referrals for other potential respondents and/or parties of interest.

Because of the diversity of interviewees and the complexity of the issues, it was not appropriate to create a standardized questionnaire. Instead, interviews were guided by a list of points that were tailored to the specific role of the informant via the spill event. Again, 35 interviews were tape recorded, transcribed, and systematically analyzed for their content[2] (as were a handful of email correspondences).

Interviews for this research are best conceived of as "conversations" in which a loose interview protocol was employed. This style of interview did not constrain the participants (or the interviewer) in ways that could have stifled personal observations or the development of tangential thoughts. Furthermore, it did not inhibit the participants from expressing their views on any of the issues raised. In this way, the interviews became not merely a conduit through which respondents could relate their experiences with the Guadalupe spill; they also facilitated responses that may have otherwise remained nascent and unrecognized.

In general, I met with little resistance when requesting interviews with regulators involved in overseeing the Guadalupe spills or with community advocates, beach users, or surfers. One major constraint was distance: the city of San Luis Obispo, the county seat, is 120 miles north of Santa Barbara, where I resided at the time. Another research constraint involved legal liability. Most potential respondents who represented the legal side (e.g., the Attorney General of California and Unocal's lawyers) as well as both current and former Unocal employees were largely unwilling to speak about the spill because of potential liability.[3]

Ethnographic Contexts

Outside of interviews, I had extensive informal contact with individuals involved with the spill through site tours, public forums, and email exchanges. In the case of site tours, local residents were invited on two occasions to tour the spill site with county officials and Unocal managers. I attended three half-day tours, and in the process I spoke with a broad range of people. Lawyers for all sides of the issue attended and were informally approachable.[4] Journalists, dunes advocates, bird enthusiasts, anti-oil advocates, unaffiliated community members, county officials, Unocal chaperons, and consultant specialists representing different interests were also present (some 40–50 people in all). Having established contact, a handful of these persons continued to keep me abreast of their "interests" via email.

I also visited the beach and the dunes intermittently over the 2½ years that this research was conducted. On those visits, I spoke informally with persons at the beach who frequented the area as surfers, fishermen, beach walkers, and for general recreation.

Moreover, because issues concerning oil are locally lumped together—whether it is resistance to offshore oil or anger about one of a number of spills (see chapter 5)—attending public events or collecting archival materials that concerned the local oil industry often led to impromptu discussions about the Guadalupe spill. Through this and another research endeavor (Beamish et al. 1998), this tendency was amplified through my identification with issues of "oil." On many occasions, spontaneous conversations also occurred in hallways and waiting rooms of offices as well as the homes of those I intended to interview with individuals I had not originally contacted or planned to meet. Although comments were not tape recorded in these instances (as they were in the more formal interviews), these conversations should not be seen as any less important than the others. New questions, better understandings, and clarity were part and parcel of asking (and in some cases answering) questions of different persons with different experiences, different backgrounds, and different interests.

Archival Materials

Along with interviews and ethnographic participation, I collected hundreds of pages of archival materials that included inter-agency and intra-agency communiqués, executive spill/remediation summaries, court documents, meeting minutes, scientific documents that characterized the extent and potential effects of the spill, official correspondences, promotional materials—both grass roots and corporate—as well as corporate memos addressed to various official parties. These documents generally fell into three categories based on their origins: government documents, environmental advocacy documents, and corporate documents. One of the strengths of thoroughly analyzing such material is that the data become a point of reference and corroboration from which to compare interviewee comments. Such documentation also revealed many of the assumptions that organizations, as corporate bodies, exhibited in relation to the spill, spills in general, environmental management, and so forth.

The Print Media

The fourth major component of my investigation was that of media presentation. Utilizing regional media outlets and a handful of national stories that covered the spill and paying close attention to how the issue was framed and discursively connected to other issues and events, I reconstructed how the public was informed of this massive pollution event. To this end, I systematically examined the print media[5] from San Luis Obispo and Santa Barbara (with special attention to the foremost regional daily in San Luis Obispo County, the Telegram-Tribune). I also examined national media using MELVYL News. MELVYL allows retrieval, by key word, of headlines in five major US newspapers: the *New York Times,* the *Los Angeles Times,* the *Washington Post,* the *Wall Street Journal,* and the *Christian Science Monitor.* This allowed for identification of the range of coverage in national media.

Media supportive of strict environmental enforcement can generate community attention and reaction. Conversely, media that portray such an event as "taken care of" or "under professional control" can take such an issue out of the public discourse. At minimum, the media help shape the context in which individuals form their positions and in which agencies take their actions.

In sum, taken as a whole, the investigation was a processual one—beginning with general understanding of what had occurred and progressively becoming more pointed as particular themes became apparent. This process of theme identification was interactive among all the components of the research. Analysis progressively evolved, as inductive research often does, to clarify recurring themes as they were uncovered through time.

Appendix B
Event Chronology of the Guadalupe Dunes Spill, 1931–1999

1931
Unocal purchases 51 percent interest in the Guadalupe oil field.

early 1950s
Diluent is developed and begins to be used at oil field.

1953
Unocal purchases the remaining 49 percent interest, becoming the sole operator of the field.

1973
An amendment to the federal Clean Water Act (section 311(b)(5) of the federal Water Pollution Control Act) institutes responsibility of industry to self-report spills of more than a barrel, making Unocal liable for excess back to this date.

1982
Concern with petroleum production near the beach is noted by regulators as early as 1982 in California Coastal Commission records.

1985
As early as 1985, Unocal records revealed in 1993, field workers noted losing 269 barrels of diluent (11,298 gallons) at the field over 17 days—a loss that was not reported.

1986
Local managers and field personnel deliberately cover up a large leak. A slick reportedly forms on the ocean bordering the site.

1988

A petroleum slick again forms on the beach and on the ocean. Local Unocal managers report the incident to the Regional Water Quality Control Board, but deny their operations as source of oil. No investigation ensues.

1990

January: A plume of diluent located under the Leroy 5X well area was identified as seeping diluent into to bordering marine environment.

February: Unocal shuts down production at the field.

February: A whistleblower calls the California Office of Emergency Services prompting California Fish and Game wardens to visit the site. They discover diluent on the beach and in the ocean.

February-March: Unocal installs a 1000-foot-long, 2-foot-wide, 18-foot-deep bentonite wall to stop diluent migration into ocean. Unocal capitulates to California Fish and Game officials and installs nine groundwater extraction wells, fifteen contamination-extraction wells, and appurtenant facilities to extract spilled thinner. In first 6 months, 390,600 gallons are recovered.

1991

January: Unocal's local supervisor submits report to San Luis Obispo County Planning Department on field activity that admits to 15 total gallons spilled since 1973.

November: The C-12 well area, adjacent to the Santa Maria River, is found to have a migrating plume of diluent. Under an emergency permit, Unocal installs a 275-foot-long PVC membrane to stop the diluent's advance toward the river. Concurrent with beach excavations, the original bentonite wall is replaced with a high-density polyethylene barrier wall (1000 feet long and 18 or more feet deep).

1992

March: A California Department of Fish and Game Office of Oil Spill Prevention and Response warden estimates that there are at least 1500 gallons of diluent on the beach sands alone.

June-July: A second whistleblower, a disgruntled worker, telephones the California Department of Fish and Game, saying he "could no longer remain silent" about Unocal's deliberate coverup of the spills.

July: A raid by government officials, based on an anonymous tip to the California Department of Fish and Game's Office of Oil Spill Prevention and Response, leads to a mountain of incriminating evidence, including boxes of files and a cache of computer disks outlining a history of unreported spills.

1993

May: Enormous petroleum plumes are discovered under a tank farm area in the inner field, away from the beach. The extent of the contamination dumbfounds officials.

June-July: In a second raid of Unocal's local offices, 67 more boxes of files and plume maps delineating the extent of the field's contamination are recovered. Twenty-eight criminal charges (all misdemeanors) are filed in district court against six Unocal employees.

December: A district court judge dismisses the misdemeanor charges against Unocal and its local field managers on grounds that the statute of limitations had run out.

1994

January: San Luis Obispo County's District Attorney files 36 new charges against Unocal, claiming that the judge misruled in dismissing previous case.

February: Extraction pumps at the site have removed 672,000 gallons of diluent.

March: The county's case against Unocal is settled through a plea bargain in which Unocal admits no guilt but agrees to pay $1.5 million in damages and is told that it is responsible for cleanup. All charges against the six employees are dropped.

March: The state of California files a civil suit against Unocal for failing to report the spills. The Surfers Environmental Alliance files a class-action suit against Unocal that accompanies the state's case.

August: On the basis of reports of repeated ocean releases, emergency permits are granted and the beach is dug up. 250,000 gallons of free-phase diluent are removed, and the sand is thermally disorbed (i.e., cooked in industrial ovens at 800–1000° to remove petroleum).

1995

February: A change of course by the Santa Maria River threatens the HDPE wall. To protect that wall, installation of a steel wall is begun.

November: Leroy 5X area wellhead number 2 and its sump (covering an acre) begin leaking into the ocean and into the Santa Maria Lagoon area

800 feet south of the larger (6 acres) Leroy 5X area plume retained by the HDPE plastic wall. 2849 cubic yards of contaminated sand are removed and stockpiled for treatment at a later date.

1996

October-November: Phase 1 of the "sheetpile" steel wall is extended 1033 feet to the north and 450 feet to the southwest, making the total length of the wall 1853 feet. The wall is further reinforced as the river deeply erodes its "impermeability." The entire project was initiated and completed under the emergency permit. The coastal commission permit specifies a two-year limit on emergency permits. Unocal estimates that the wall will have to remain in place for 15 years—the time it will take to remove contaminated sands by excavation and other techniques.

1997

October: Unocal estimates that 8.5 million gallons have been spilled, yet records show that 20 million gallons of diluent remain unaccounted for. If this is the case, it is the largest spill in US history.

November: The 7X area plume (adjacent to the 5X beach area) is dredged and an 18-foot-high berm is created in preparation for high tides and groundwater levels associated with an El Niño event. Some 14,000 cubic yards of sand are removed. The 33,000 cubic yards used to create the protective berm are then used to fill the football-field-size hole in the beach.

1998

February: the 5X plume continues to seep into the ocean and the river despite attempts by Unocal and by regulatory agencies to stop it with walls and early-winter excavations. The Santa Maria Flood Control District releases water from the Twitchell Reservoir. The raging river cuts into and releases oil from an old sump, releases more diluent from an underground plume, and releases asphalt chunks from an old road paved with petroleum products.

July: Out-of-court settlement of the state's civil case against Unocal nets $43.8 million.

1999

County and state officials endorse the environmental impact report and Unocal's cleanup strategy. The plan calls for two 10-year plans.

Oil sumps containing hundreds of thousands of gallons and (more ominous) PCB contamination are found in the interior of the oil field.

Appendix C
Regulators and Regulations Involved in the Guadalupe Spill

Institutional Participants

(Asterisks indicate agencies that have held leadership roles in investigating, interdicting, and remediating the spill.)

United States
Coast Guard*
Army Corps of Engineers
Environmental Protection Agency
Fish and Wildlife Service
National Marine Fisheries Service
National Oceanic and Atmospheric Administration

California
Air Resources Board—Compliance Division
Environmental Protection Agency—Department of Toxic Substances Control
Coastal Commission
Coastal Conservancy
Department of Conservancy—Division of Oil, Gas, and Geothermal Resources
Department of Fish and Game*—(post-1990) Oil Spill Prevention and Response Unit
Environmental Health Hazards Assessment
Office of Emergency Services
Regional Water Quality Control Board*
State Land Commission
Toxic Substances Control

San Luis Obispo County
Air Pollution Control District
Department of Agriculture
Department of Health
Department of Planning and Building*

Santa Barbara County
Energy Division

Others
Chumash Nation
Native American Heritage Commission

Mission Statements

Federal Regulatory Agencies

National Marine Fisheries Service
The Protected Species Management Division of the NMFS is responsible
for the management of protected marine species (such as marine mammals,
sea turtles, and Chinook salmon) under the provisions as set out by the
Endangered Species Act and the Marine Mammals Protection Act. The
PSMD reviews National Environmental Protection Act and California
Environmental Quality Act environmental documents prepared for pro-
jects that could affect protected marine species.

Army Corps of Engineers
The Army Corps of Engineers has regulated certain activities in the nation's
waterways since 1899. The regulatory jurisdiction of ACOE includes all ocean
and coastal waters within the zone 3 nautical miles seaward of territorial
seas. Wider zones are recognized in the navigable waters of the United States.

Coast Guard
As the primary overseer of navigable US waters, the Coast Guard is involved
with a number of issues which overlap oil industry activity. First of these is
the delineation of navigable waters. The Coast Guard requires aids to navigation on artificial reefs and other fixed structures such as offshore platforms.

Furthermore, the Coast Guard is involved in the control of pollution by oil and other hazardous substances, and in the removal of such discharge(s). The Coast Guard is notified and takes a lead role in the advent of marine release. More recently, the Oil Pollution and Prevention Act of 1990 has further specified the Coast Guard's role in the prevention, response, and cleanup, giving the Coast Guard a greater responsibility to direct emergency reactions to oil spill events.

Environmental Protection Agency
The primary concern of the this agency with regard to on and offshore oil development (and currently its decommissioning) is through the Clean Water Act. Through the National Pollution Discharge Elimination System Permit Program, the EPA regulates the discharge (from runoff to systems "flushing") of effluent from industry infrastructure—pipelines, platforms, and onshore facilities.

Fish and Wildlife Service
This service's primary concern is the protection of public fish and wildlife resources and their habitats. It mandates require that it provide comments on any public notice issued for a federal permit or license affecting the nation's waterways, in particular, COE permits pursuant to section 404 of the Clean Water Act and section 10 of the Rivers and Harbors Act of 1899. Additionally, the FWS administers certain amendments of the Endangered Species Act of 1973.

Federal Environmental Policy

Clean Air Act of 1967
Originally providing for research and training in air pollution control techniques and assessment, this act was given "teeth" by amendments made in 1967, 1970, and 1977. In 1967, the federal government set ambient standards for air pollutants in addition to sponsoring state initiatives to assure federal standards were met. Ten years later saw the federal government substantially increased its enforcement role and set stricter guidelines for (primarily vehicular) combustion emission standards. In the most ambitious 1977 amendment, the federal government required states to adopt plans for full compliance with federal standards by 1982.

Endangered Species Act of 1969

Section 9 of this act prohibits the "take" of any listed specie, which means to harass, harm, hunt, shoot, wound, kill, trap, capture, or collect or attempt to engage in any such conduct. A notable component of this act is the inclusive definition of "harm" which it operationalizes. Harm includes significant habitat modification or degradation, scenarios that significantly impair essential behavioral patterns, including breeding, feeding, or sheltering. Any individual or organization that "takes" (that is, harms) listed wildlife (and, by extension, their habitats) is subject to prosecution under section 9 or section 10.

Other sections specify that all federal agencies use their authorities in the furtherance of the purposes as laid out by the ESA by carrying out programs for the conservation of endangered species and threatened species. Furthermore, the federal government is mandated by the act to review proposed activities which may affect listed species.

Federal Coastal Zone Management Act of 1972

The Coastal Zone Management Act of 1972 stipulates that federal agencies, in carrying out their functions and responsibilities with regard to coastal resources, consult, cooperate, and coordinate their activities with public, state, and regional authorities in the development of coastal management plans. After approval of a state's coastal management program, applicants for federal licenses or permits to conduct an activity affecting land or water uses in the coastal zone are required to provide certification that the proposed activity complies with that state's approved coastal program, and that the proposed activity will be conducted in a manner consistent with that program.

After the management program of any coastal state has been approved by the Secretary, the act also calls on applicants to furnish to the state or other designated agency a copy of that certification, with all necessary information and data included. Any person who submits a plan for the exploration or development of, or production from, any area which as been leased under the Outer Continental Shelf Lands Act (43 USC 1331 et seq.) is required to attach to certification that each activity complies with that state's approved management program and will be carried out in a manner consistent with such program. No federal official or agency shall grant such person any license or permit for any activity until that state or its designated agency receives a copy of such certification and plan, together with any other necessary data and information.

Federal Water Pollution Control Act of 1948

Intended to establish water quality standards for coastal waters, this act also set procedures for the removal and cleanup of discharged oil and other water-borne effluents. It also provides guidelines for the cost reimbursement of cleanup (liability) of hazardous materials discharge. Under the FWPCA, the Coast Guard is the lead agency in marine environmental response, port environmental safety, and (marine) waterway management.

Specifically, Amendment P.L. 92-500 to the FWPCA prohibits the unauthorized discharge of dredged or fill material into United States waters. The selection and use of dredged disposal sites will be in accordance with the guidelines developed by the Environmental Protection Agency. The EPA can deny, prohibit, restrict, or withdraw the use of any defined area as a disposal site whenever (if it is determined, after notice and opportunity for public hearing and after consultation with other relevant agencies) that discharge of such materials into such areas will have an unacceptable adverse effect on municipal water supplies, fishery areas, wildlife, or recreational areas. This act is administered in conjunction with the Army Corps of Engineers.

National Environmental Protection Act of 1969

Enacted in January of 1970, this requires all administrative agencies of the federal government to consider the environmental impacts of their actions in the process of project development and decision making. NEPA also allows other officials, Congress, and the public to independently evaluate the environmental consequences of government actions.

Through section 102 NEPA also requires that environmental impact statements be produced for all federal actions that could affect the environment. These statements must address the environmental impact of the proposed action(s), any adverse environmental affects that cannot be avoided should the proposal be implemented, alternatives to the proposal, the relationship between local short-term uses of the environment and the maintenance and enhancement of long-term productivity, and any irreversible and irretrievable commitments of resources involved in the proposal. The primary purpose of the EIS process is to ensure that the policies and goals of the NEPA are carried out. Thus, federal agencies are to base their decisions on information found in the EIS and other materials.

Whether an EIS is carried out or not depends on if it is required, which in turn is based on whether the proposal under consideration constitutes a major federal action that will significantly affect the environment. Federal action means not only those the federal government undertakes, but also those it permits or approves. The standard "significantly affecting the quality of the human environment," means having an important or meaningful affect upon a broad range of aspects of the human environment.

The basic rules for determining whether a EIS is adequate are whether the agency in good faith has taken an objective look at the environmental consequences of the proposed action and alternatives, whether the EIS provides detail sufficient to allow those who did not participate in its preparation to understand and consider the pertinent environmental influences involved, and whether the EIS explanation of alternatives is sufficient to permit a reasoned choice among different courses of action.

National Marine Fisheries Enhancement Act of 1972
See agency manifesto above.

Oil Pollution Act of 1990
This act introduced new provisions for oil pollution liability, prevention, preparedness, and cleanup pertaining to vessels, offshore oil and gas facilities, onshore terminals, and other petroleum industries. Major provisions of the law include oil pollution liability and compensation, prevention and removal of oil pollution, oil pollution research and development program, and amendments to the oil spill liability trust fund. The Coast Guard has been given (by this act) greater responsibility for directing emergency response to marine oil spills.

Federal Assessment Strategies

Natural Resources Damage Assessment
This act guides personnel involved in emergency response to oil spills and hazardous substance releases. Under the National Oil and Hazardous Substances Pollution Contingency Plan the NRDA process provides a protocol template. Federal on-scene coordinators and other (federal and state; see below) natural resource trustees are called on to work with one another on projects which cross jurisdictions as is the rule with oil spills. NCP directs on-scene coordinators to work with trustees in specific preparedness and response activities, to help ensure that natural resources are protected when they are at risk from an actual or potential oil spill or hazardous substance release.

In particular the NRDA process involves the coordination of assessment activities between response operations in order to assure that data from the multiple activities that take place in such response scenarios can more effectively support those (referred to as a on-scene coordinator) in assessing the damages incurred at the site of the spill. Trustees are federal officials designated by the president; state officials designated by the governor; Indian officials designated by the governing body of any Indian tribe; and or foreign officials designated by the head of any foreign government; each acts on behalf of the public (of the nation, the state, the tribe, or the foreign country). The purpose of an NRDA assessment is to: avoid or minimize injury to natural resources; assess damages or injury to, destruction of, or loss of natural resources; obtain compensation from the responsible party for any damages done through negotiation or litigation; and develop and implement plans for restoration of damages or injured resources.

Environmental Impact Statement
See National Environmental Protection Act of 1969 above.

California State Regulatory Agencies

California Air Resources Board
The Air Resources Board's mission is to promote and protect public health, welfare and ecological resources through the effective and efficient reduction of air pollutants in recognition and consideration of the effects on the economy of the state. (See Regional Regulators and Policies below for county Air Pollution Control Districts.)

California Coastal Commission
The California Coastal Act of 1976, the foundation for the federally approved California Coastal Management Plan, was enacted by the state legislature to provide for the conservation and development of the state's 1,100 mile Coastline. Under the Coastal Act and the CCMP, the commission must consider the impacts of proposed projects within the coastal zone.

California Department of Fish and Game
The primary responsibility of this department in regard to oil and oil-related developments is to review NEPA and CEQA documents with respect to fish

and wildlife resources and habitat impacts resulting from project implementation. California's Oil Spill Prevention and Response Act of 1991 was a direct response to the threat posed by an *Exxon Valdez* or Huntington Beach type tanker spill off the California Coast. OSPR, within the Department Fish and Game, is the lead state agency charged with oil spill prevention and response within California's marine environment. The Lempert-Keene-Seastrand Oil Spill Prevention and Response Act of 1990 established OSPR and provides the OSPR Administrator with substantial authority to direct spill response, cleanup, and natural resource damage assessment activities.

California Division of Oil and Gas and Geothermal Resources
This subdivision of the California Department of Conservation is responsible for supervising the drilling, operation, maintenance, and abandonment of wells throughout the state, including those wells within territorial seas. Division inspectors conduct on site inspections to ensure compliance with DOGGR regulations.

California Environmental Protection Agency
The mission of this agency is to improve environmental quality in order to protect public health, the welfare of the state's citizens, and California's natural resources. Many agencies fall under the "umbrella" provided by this agency. Their jurisdictions overlap portions of the oil production process. The agencies include: the air resources board, department of toxic substances control, integrated waste management board, office of environmental health hazard assessment, state water resources control board.

California State Lands Commission
This commission is responsible for the management of extractive development of mineral resources located on state lands. Oil and gas development has primarily been concentrated on sovereign tide and submerged state lands adjacent to the coast and out 3 nautical miles offshore of Southern California.

California State Office of Emergency Services
Under the authority of the Emergency Services Act, this office mitigates, responds to, and aids in recovery from the effects of emergencies that threaten lives, property, and the environment. The state provides a pivotal link in disaster management by assisting local governments with response

and recovery. OES oversees the California Mutual Aid system, is responsible for the Operational Area Satellite Information System, leads in the Standardized Emergency Management System, and is the proponent agency for other technical programs, particularly in the radiological and hazardous materials areas. During disaster response, OES coordinates the activities of state agencies. When necessary, OES recommends that the governor proclaim a disaster and, if warranted, prepares the petition the governor uses to request a Federal Declaration. OES also implements the state's Natural Disaster Assistance Act, which provides recovery funding for local governments suffering disaster losses. Additionally, OES coordinates all federal disaster activities in the state, ranging from hazard mitigation to response and recovery.

California Department of Toxic Substances Control
The mission of this department is protection of public health and the environment through effective and efficient regulation of hazardous waste management and site mitigation activities and through promoting the development and use of pollution prevention and waste minimization technologies. In accomplishing this mission, the DTSC is committed to carrying out all program activities in a manner that is responsive to the public and to industry needs. DTSC has primary authority over disposal of hazardous waste and toxic wastes. DTSC administers both the Subtitle "C" of the federal Resource Conservation and Recovery Act and other applicable California Toxic Substance laws.

California Office of Environmental Health Hazards Assessment
The mission of this office is to protect and enhance public health and the environment by objective scientific evaluation of risks posed by hazardous substances. This is the lead agency for the laws established by the state's Proposition 65, which prohibit the release into drinking water of suspected carcinogens and reproductive toxins. Some by-products of petroleum production process, such as benzene, toluene, xylene, are listed as suspected carcinogens.

California Water Resources Control Board
The mission of this board, created by the legislature in 1967, is to preserve and enhance the quality of California's water resources and to ensure their proper allocation and efficient use for the benefit of present and future

generations. Additionally, the state board ensures the highest reasonable quality of waters of the state, while allocating those waters to achieve the optimum balance of beneficial uses. The joint authority of water allocation and water quality protection enables the state board to provide comprehensive protection for California's waters.

The board is responsible for designated non-hazardous and inert wastes that may enter state waters. Drilling muds, tank bottom wastes, and oil-contaminated water may become designated waste subject to state water quality regulations.

California State Environmental Policies

Air Toxics "Hot Spots Act"
This program regulates 720 substances through an agenda that requires facilities to inventory emissions, asses health risks from those emission, identify "hot spots," advise nearby populations of them, and reduce significant associated risks. Local air districts implement this program, usually by applying a ten parts per million criteria risk standard.

California Coastal Act of 1976
This act created a unique partnership between the state (acting through the California Coastal Commission) and local government (15 coastal counties and 58 cities) to manage the conservation and development of coastal resources through a comprehensive planning and regulatory program. The 1976 Act made permanent the coastal protection program launched on a temporary basis by a citizens' initiative that California voters approved in November of 1972 (Proposition 20, the "Coastal Conservation Initiative"). The Act's coastal resources management policies and governance structure are based on recommendations contained in the California Coastal Plan called for by Proposition 20 and adopted by the Coastal Commission in 1975 after 3 years of planning and hundreds of public hearings held throughout the state. The CCA, the foundation for the federally approved California Coastal Management Plan, was enacted by the state legislature to provide for the conservation and development of the state's 1,100-mile coastline. Under the CCA and CCMP, the Coastal Commission (see above) must consider the impacts of development in all its forms including the inverse—removal and abandonment procedures.

California Environmental Quality Act

Enacted in 1970, this was a response to growing concern about environmental protection and has four basic purposes: to inform the public and governmental decision makers of potential environmental effects of proposed activities; to identify ways to reduce or avoid environmental damage; to prevent damage by requiring changes in projects through alternative projects of mitigation measures; and to make the public aware if an approved project will have significant environmental effects.

CEQA applies to any activity proposed, funded, or permitted by a state or local agency that has the potential for resulting in physical change in the environment. Only projects statutorily or categorically exempt from CEQA review are exempt. Otherwise, projects which fall within these parameters must prepare either a Negative Declaration or Environmental Impact Report to assess potential impacts to the environment. Generally, an EIR is required when a project has the potential for significant environmental impact; ND is prepared when there is considerable evidence that a project will cause no substantial effect on the environment.

CEQA calls for not only permitting but the continued monitoring of permitted projects. An agency must adopt a reporting and monitoring program whenever it makes a finding relevant to the mitigation or avoidance of significant environmental effect of a project.

The CEQA process, as it is referred to, normally consists of three parts or phases. The first phase consists of a preliminary review of a project to determine whether it is subject to CEQA. The second involves preparation of an initial study to determine whether the project may have a significant environmental effect (if not, a negative declaration). The third phase is the preparation of an EIR if the project is determined to have significant effects.

Lempert-Keene-Seastrand Oil Spill Prevention and Response Act

This act provides civil and criminal penalties for knowingly, intentionally or negligently discharging or spilling oil into oceans and other marine waters by specified owners and operators of marine facilities and vessels used for the production, processing or transport of oil (Government Code, sections 8670.64–8670.67). However, OSPRA does not apply to inland spills, where most spills occur from refineries, railroad and vessel, and ruptured pipelines.

Toxic Air Contaminants Identification and Control Act

This law requires the California Air Resources Board to identify toxic air contaminants, assess risks, and then, if necessary, to develop methods to eliminate the contamination or reduce the health risk. Local APCDs implement the control measures.

California State Assessment Strategies: Environmental Impact Report

See California Environmental Quality Act above.

Regional Environmental Regulators

County of San Luis Obispo Air Pollution Control District

The San Luis Obispo Air Pollution Control District monitors and enforces air pollution standards. The support and cooperation of the public and business community are crucial to the success of their efforts. Besides being vital to public health, clean air is critical to two of the county's most important industries: agriculture and tourism.

The San Luis Obispo APCD was formed in 1970 and is a state district created by state law. The district's staff of 20 includes engineers, inspectors, planners, technicians, and administrative personnel. It is the basic policy of the Air Pollution Control Board and the APCD to control emissions of air contaminants within district boundaries, so as to achieve and maintain state and federal ambient air quality standards. This, in turn, promotes and protects public health, public welfare and the productive capacity of the citizens of San Luis Obispo.

The APCD has the primary responsibility for controlling emissions from mobile and stationary sources of air pollution. Sources vary, from power plants and refineries, corner gas stations and dry cleaners, to personal autos and commercial cargo service. The APCD planning staff is also responsible for evaluating emission control measures that may be needed to protect and improve local air quality. Planners review new residential, commercial and industrial projects and develop strategies to minimize their air quality impacts. APCD engineers evaluate plans and issue permits for any new project that involves installing, altering, or operating equipment that either causes air pollution or is used to control it. Engineers work with permit applicants to minimize emissions of air contaminants and to ensure compliance with all federal, state, and local rules and regulations. Once a new facility is completed and begins operation, APCD inspectors conduct periodic inspections to ensure compliance with permit requirements. Along with evaluation and permitting the APCD can also takes enforcement action to bring businesses into compliance.

County of Santa Barbara Air Pollution Control District
See County of San Luis Obispo Air Pollution Control District above.

Regional Water Quality Control Boards
See California Water Resources Control Board above. As regional representative of the State Water Resources Control Board, the mission of the regional board is to ensure the highest reasonable quality of waters of their regions, while allocating those waters to achieve the optimum balance of beneficial uses.

County of San Luis Obispo, Department of Planning and Building, Energy and Natural Resources Division
The Energy and Natural Resources Division of the Department of Planning & Building is responsible for the planning and project review of all proposed energy and mining development within San Luis Obispo County. This includes the processing of all land use permit applications for onshore energy and surface mining projects; the monitoring and evaluation of all offshore oil development studies; the development of proposals for submittal to the state and federal governments for protection of waters off the county coast; and the coordination and development of an ongoing inspection program for all surface mining operations. This includes special attention to environmental review. The division also assists elected and appointed county officials and the public in evaluating potential impacts of land use projects, both private and county, on environmental resources as required by the California Public Resources Code.

County of Santa Barbara Planning and Development Department, Energy Division
The Energy Division works to influence federal and state energy policy in the interests of the citizens of Santa Barbara County. Important tasks which the Energy Division pays particular attention include development of local energy plans; promulgation of policies and ordinances to best meet adopted

federal, state and local goals; participation in joint federal, state and local review panels for the environmental review and permitting of major oil and gas development projects; and review of oil and gas projects and with the permit conditions imposed by the County decision makers.

Regional Environmental Assessment Strategies

Environmental Impact Reports
Done in conjunction with state of California; see above.

Local Coastal Plans
See Federal and California Coastal Acts above.

Environmental Laws and Regulations Governing Oil and Gas Exploration, Production, and Accidents

The following is a summary of the regulatory "domains" the petroleum industry must address when producing oil in the San Luis Obispo County region. Functionally, regulated activities are broken into nine general categories:

Produced Water Management
Waste Management
Emergency Preparedness and Response
Land Access, Land Use, and Endangered Species
Air Quality
Toxic Air Contaminants
Hazardous Materials Handling and Storage
Transportation and Pipelines
Oil Spill Prevention and Response

Each of these nine areas is detailed further below. After the summaries of regulated areas and how they correspond to spilled oil, I list all the agencies that have been involved with the spill and describe each agency's public trustee responsibilities and relation to the spill site. I also list the principal environmental acts in place during the spillage.

Management of Produced Water
"Produced water" is water taken out of the ground along with petroleum. California's oil activities generate 2.5 million barrels annually. That is nearly

seven barrels of water for every barrel of oil produces (State of California, Department of Conservation, Division of Oil & Gas, 1986). Produced water can contain a number of chemicals which would make disposal a significant problem. (See table C.1.)

The regulation of produced water crosses a number of California state jurisdictions, depending on how the water is being treated, where the water ends up (re-injected into wells, into local sewer systems, or in standing ponds), and if an accidental discharge has taken place. The oversight agencies include the California Division of Oil, Gas, and Geothermal Resources;

Table C.1
Laws and regulations pertaining to oil spills. Source: A Profile of California's Oil and Gas Industry (California Department of Conservation, Division of Oil & Gas, 1986).

	Requirement	Responsible agencies
Federal Water Pollution Control Act, Federal Oil Spill Act	Oil spill pollution prevention, preparedness, response	US EPA, US Coast Guard, Bureau of Land Management
Clean Water Act, Porter Cologne Water Pollution Control Act, California Code of Regulations, Titles 22 and 23	Oil spill prevention, control, and counter measure plan; oil spill reporting	US EPA, US Coast Guard, State Division of Oil, gas, and Geothermal Resources, State Lands Commission, California Department of Fish and Game
Code of Federal Regulations 40 CFR 3160, California Public Resources Code, California Government Code, California Code of Regulations, Title 14	Oil spill and response; oil spill reporting	Division of Oil and Gas, State Lands Commission
California Code of Regulations title 19	Emergency response regulations; oil spill reporting	California Office of Emergency Services
California Public Resources Code, California Government Code	Marine terminal and facility oil spill prevention and response; oil spill reporting; accidental and hazards assessment	California Department of Fish and Game, State Lands Commission

the California Regional Water Quality Control Board; the California Department of Toxic Substances Control; the California Department of Fish and Game; and the US Fish and Wildlife Department.

Waste Management

In California, waste management is divided into two components: solid and hazardous wastes. Due to the limited number of landfills, concerns about the contamination of groundwater supplies, and the recent generation of a high volumes of toxic waste based in its large urban population and manufacturing sector, California's waste management laws are, generally speaking, more strict than those found in the rest of the country and than the federal governments.

The impact of these tight regulations on petroleum producers has been quite heavy. The oil and gas industry produce a million tons of solid waste annually with 140,000 tons or 14 percent of this being classified as hazardous by California standards. In comparison, by federal government standards 35,000 tons of the same material is considered hazardous. (See table C.2.)

Emergency Preparedness and Response

Fifteen federal, state, and local agencies play a part in preparing for and responding to emergencies. For the oil industry emergencies can involve oil spills, accidental release of hazardous substances, and blowout and discharge prevention and containment. The responsibilities which comprise this category of regulation are broken further into three jurisdictional classifications: Oil Spills, Hazardous Materials, and Discharges.

The prevention, reporting, and assessment of oil spills at oil fields is primarily the responsibility of the California Division of Oil, Gas, and Geothermal Resources. For all other spills, responsibility falls on the California Department of Fish and Game. In cases of spills into a body of water, the US EPA, the Coast Guard, the California State Lands Commission, the California Department of Fish and Game, and California's Regional Water Quality Control Boards exercise their responsibility to control the spills and enact countermeasures.

In the case of hazardous materials (see below), the California Department of Toxic Substances Control, the California Office of Emergency Services, and the US EPA have primary responsibility. Response to release reports, cleanup, and remedial actions and are in the main guided by three

Table C.2
Laws and regulations pertaining to toxic air contaminants. Source: A Profile of
California's Oil and Gas Industry (California Department of Conservation,
Division of Oil & Gas, 1986).

	Requirement	Responsible agencies
Federal Clean Air Act	National emissions standards for hazardous pollutants; maximum achievable control technology for federally defined hazardous air pollutants	US EPA, South Coast AQMB: County APCDs
Health and Safety Code—Toxic Air Containment	Controls on California "list of Toxic Air Contaminants"	California Air Resources Board, South Coast AQMB, County APCDs, Office of Environmental Health Hazards Assessment
Health and Safety Code—Air Toxic "Hot Spots"	Air toxic emissions inventory, risk assessment, and public notice	California Air Resources Board, South Coast AQMB, County APCDs, Office of Environmental Health Hazards Assessment
South Coast AQMB rule 1401	New source review carcinogenic pollutants	South Coast AQMB
South Coast AQMB rule 1410	Hydrogen fluoride controls	South Coast AQMB
Ventura county APCD	Chromium cooling tower controls	Ventura County APCD

federal laws: the Comprehensive Environmental Response Act, the
Compensation and Liability Act, and the Resource Conservation and
Recovery Act.

Finally, regarding petroleum discharges, the California Division of Oil,
Gas, and Geothermal Resources (in conjunction with the Bureau of Land
Management) issues drilling operations permits that include safety and discharge prevention and containment provisions.

Other agencies involved at a peripheral level include the California
Highway Patrol and the US Department of Transportation, including local
fire, health, and emergency planning departments. (See table C.3.)

Table C.3
Laws and regulations pertaining to emergency preparedness. Source: A Profile of California's Oil and Gas Industry (California Department of Conservation, Division of Oil & Gas, 1986).

	Requirement	Responsible agencies
Federal Comprehensive Environmental Response, Comprehensive and Liability Act	Hazardous substances release reporting; clean up and remedial actions; liability assurance	US EPA, California Toxic Substances Control, California Office of Emergency Services
Emergency Planning and Community Right to Know Act	Updating of material safety data sheet; inventory reporting; release Reporting	US EPA, Local Emergency Planning Committee, Local Fire and Health Agencies
California Public Resources Code, California Government Code	Spill prevention and response; spill reporting, accident, and hazards assessment	California Division of Oil, Gas, and Geothermal Resources, Office of Emergency Services
California Public Resources Code and California Code of Regulation, Title 14	Spill prevention and response; accident and hazards assessment	California Division of Oil, Gas, and Geothermal Resources, State Land Commission, Office of Emergency Services
Clean Water Act, Porter Cologne Water Pollution Control Act	Oil spill prevention, control, and countermeasure plan; spill reporting	US EPA, US Coast Guard, State Water Resources Board, Regional Water Quality Boards, California Office of Emergency Services
US Code, 49 CFR 394, Vehicle Code	Hazardous materials transport; accidental spills of waste and hazardous substances	Federal Department of Transportation, California Office of Emergency Services
Federal Pipeline Safety Act, California Pipeline Safety Act	Accidental spills from pipelines; spill reporting	State Fire Marshal, California Office of Emergency Services
Resource Conservation and Recovery Act, California Hazardous Waste Control Act	Hazardous waste emergency response; hazardous waste reporting plan	California Department of Toxic Substances Control, US EPA

California Health and Safety Code (AB 2185/2187)	Hazardous material business; emergency response plan; emergency response plan	California Office of Emergency Services, Local implementing agency
California Health and Safety Code	Risk management business; emergency response plan; H/S inventory; emergency response plans	California Department of Toxic Substances Control, Local Implementing Agency—County Health/Fire Departments
Code of Federal Regulation 40 CFR 3160.4, California Public Resources Code and California Code of Regulations	Drilling operations permit includes: safety, blowout, discharge prevention, containment and well abandonment	Bureau of Land Management, California Division of Oil, Gas, and Geothermal Resources
US Hazardous Materials Regulations, California Vehicle Code	Hazardous materials transport; emergency response and procedure; reporting	US Department of Transportation, California Highway Patrol
California Code of Regulations, Title 19	Emergency response regulations; oil spill reporting	California Office of Emergency Services
California Underground Storage Tank Act	Emergency response; regulation; reporting for releases underground storage tanks	California Regional Water Quality Control Board and/or local agency

Land Access, Permitting, and Endangered Species

County governments issue land use permits and act as lead agencies in the application of the California Environmental Quality Act, which requires potential development projects to first conduct an EIR to assess potential environmental impacts. When a development is proposed for the coast, the California Coastal Commission is lead permitting agency, and local agencies often also require building permits.

The petroleum industry must also comply with federal and California state Endangered Species Acts, administered respectively by the US Fish and Wildlife Service and the California Department of Fish and Game. (See table C.4.)

Table C.4
Laws and regulations pertaining to transport and pipelines. Source: A Profile of California's Oil and Gas Industry (California Department of Conservation, Division of Oil & Gas, 1986).

	Requirement	Responsible agencies
US Code 49 CFR 394 Vehicle Code	Hazardous material transport; accidental spills of waste or hazardous substance	Federal Department of Transportation, California Office of Emergency Services
Federal and State Pipeline Safety Act	Pipelines safety regulations control the operations and maintenance of pipelines, hydrostatic testing, and records keeping	US Department of Transportation, State Fire Marshal
Health and Safety Code, Hazardous Waste Control Act	Regulations that control the transportation of hazardous materials as well as the transport of oil	California Department of Toxic Substance Control, California Integrated Waster Management Board
California Public Resource Code: California Government Code	Marine terminal and facility controls, accidents, and hazardous waste assessment	State Lands Commission
Local land use and zoning regulations	Local land use permits	City and county planning departments

Air Quality

California air emissions laws date from 1948. Currently half of the state's air emissions come from mobile sources, which are regulated by the California Air Resource Board. The other half, emitted from stationary sources, are regulated by 34 different and semiautonomous Air Pollution Control Districts.

Federal air quality standards are enforced by the US EPA, but these standards are implemented by CARB and APCD. In many cases the standards imposed by the California Clean Air Act actually exceed those laid out in federal standards.

The oil industry's air emissions have been heavily regulated since the late 1970s, with special attention focused on sulfur dioxide and ozone emissions. (See table C.5.)

Table C.5
Laws and regulations pertaining to produced water. Source: A Profile of California's Oil and Gas Industry (California Department of Conservation, Division of Oil & Gas, 1986).

	Requirement	Responsible agencies
Federal Safe Drinking Water Act	Injection well permits and controls	California Division of Oil, Gas, and Geothermal (DOGGR), Federal BLM, US EPA
California Public Resources Code & California Code of Regulations Title 14 & Title 23, California Toxic Pits Control Act, Fish and Game Code	Surface impoundment controls	DOGGR, State Water Resource Control Board, US EPA, Regional Water Quality Control Boards, California Department of Fish and Game
California Public Resources Code & California Code of Regulations Title 14 & Title 23, California Toxic Pits Control Act, Fish and Game Code	Surface Water Discharge Permit process: NPDES permit and discharge waster requirements; NPDES and hazardous substance reporting; liability financial assurance	California Division of Oil, Gas, and Geothermal (DOGGR), Federal BLM, US EPA, USCG, Office of Emergency Services
California Health and Safety, Porter Cologne Water Pollution Control Act, California Code of Regulations, Title 23	Controls all permits for water treatment facilities that may generate California only hazardous waste	California Office of Toxic Substances Control
Proposition 65	Warning of public on discharge required; prohibits discharge into drinking water	California Division of Oil, Gas, and Geothermal (DOGGR), Federal BLM, US EPA, USCG, Office of Emergency Services

Toxic Air Contaminants

Toxic air contaminants differ from air pollutants such as smog in that they may contain potential carcinogens and pose other health risks. The federal Clean Air Act (1967) sets national emission standards for hazardous air pollutants, identifies 190 such substances, and imposes control technology to limit their emission.

The State of California has enacted two laws to regulate Toxic Air Contaminants: the Toxic Air Contaminants Identification and Control Act and the Air Toxics Hot Spots Act. Nearly 400 substances are identified as high risk. Oil and gas production typically involve 40 of these substances.

The Air Toxics Program is similar to Proposition 65[1] in that it calls on regulators to release public warnings if cancer causing agents are emitted into the air. (See table C.6.)

Hazardous Materials

In addition to the approximately 140,000 tons of hazardous waste oil and gas producers must transport and dispose of, the industry also uses a significant volume of hazardous of materials in their production processes. The US Department of Transportation, the California Office of Emergency Services, and the California Highway Patrol regulate the transport of these materials and have oversight responsibility in the event of accidental release.

The State and Regional Water Quality Control Boards regulate both under and above-ground storage tanks for hazardous materials. All such facilities must also meet federal and state Occupation Health and Safety Standards. California in particular is unique in its warning requirements. Under Proposition 65 and other right-to-know laws, facilities must make publicly available—or face legal action—information about the safety of stored materials, the use of chemicals on site, and the release of any hazardous materials used. (See table C.7.)

Transportation and Storage

As many as ten governmental agencies oversee this portion of the petroleum industry. The US Department of Transportation and State Fire Marshal oversee pipeline safety, the State Lands Commission oversees marine terminals, and city and county planning departments apply regional standards and accompanying permits to their areas.

Table C.6
Laws and regulations pertaining to waste management. Source: A Profile of
California's Oil and Gas Industry (California Department of Conservation, Division
of Oil & Gas, 1986).

	Requirement	Responsible agencies
Resource Conservation and Recovery Act, California Hazardous Waste Control Act, California Code of Regulation, Title 22	Solid waste management controls, Hazardous waste management controls: H/W treatment, storage and disposal facility permits and controls; cleanup and remedial actions; H/W transport controls; H/W sources reduction plans; biennial report	California Department of Toxic Substances Control, US EPA, California Integrated Waste Management Board
California Integrated Waste Management	Solid waste management controls; disposal permits	California Integrated Waste Management Board: Local enforcement agencies
Federal Comprehensive Environmental Response, Comprehensive Liability Act	Clean up and Remedial Actions; liability assurances	US EPA, California Department of Toxic Substance Control
Toxic Substances Control Act	PCB waste controls; asbestos waste controls	US EPA, California Department of Toxic Substance Control
California Hazardous Waste Control Act	Used oil recycling	California Department of Toxic Substance Control, California Integrated Waste Management Board
Clean Water Act dredge and fill regulations	Permits for dredging and fill disposal in US waters	US Army Corps of Engineers, Regional Water Quality Control Boards
County Sanitation and Public Works Department Regulations	Industrial waste disposal permits and controls	County Sanitation and Public Works Departments
Local land use permits	Land use permits	Cities, counties

Table C.7
Laws and regulations pertaining to land access and land use. Source: A Profile of California's Oil and Gas Industry (California Department of Conservation, Division of Oil & Gas, 1986).

	Requirement	Responsible agencies
NEPA, Public Resources Code, California Environmental Quality Act	State environmental quality regulations including the use of environmental impact reports to a asses and mitigate environmental impacts; NEPA and its EIS process on federal lands	Lead agency is designated, usually a city or county planning agency." All other agencies are involved through EIR/EIS process. On federal Land BLM or DOE is lead.
Endangered Species Act	Land-use permit for any activity on public land that could affect listed specie	US Fish and Wildlife, California Department of Fish and Game
Federal Land Policy and Management Act	Right-of-way permits for projects on public lands; Construction, operations and rehabilitation plans	Bureau of Land Management
Public Resources Code, Coastal Management Act	Coastal development permits for facilities in the sate coastal zone	California Coastal Commission
Local planning, zoning, land use, and building ordinances	Local land-use and building permits	City and county planning departments and agencies

Transportation of hazardous materials falls at the federal level under the jurisdiction of the US Department of Transportation and at the state level under the Office of Emergency Services. Transportation of used oil and hazardous waste may also involve regulations enforced by the California Highway Patrol, the California Department of Toxic Substances, and the California Integrated Waste Management Board. Local fire and emergency planning departments may also be involved in preventing and responding to emergency spills. (See table C.8.)

Oil Spill Prevention and Response
California's Lempert-Keene-Seastrand Oil Spill Prevention and Response Act of 1991 was to a great degree a response to the *Exxon Valdez* tanker

Table C.8
Laws and regulations pertaining to air quality. Source: A Profile of California's Oil and Gas Industry (California Department of Conservation, Division of Oil & Gas, 1986).

	Requirement	Responsible agencies
Federal Clean Air Act	Air pollution control, including air quality standards; emissions controls and reporting; permitting; new source review; prevention of significant deterioration	US EPA, California Air Resources Board, Regional Air Pollution Control Districts
California Health and Safety Code, California Clean Air Act	Air emissions reductions mandates controls on air districts SIPs	US EPA, California Air Resources Board, Regional Air Pollution Control Districts
South Coast AQMD rules	New and modified source review, permit to construct, new source performance standards, stack monitoring, source sampling and testing, so controls, no controls, solvent and rvp controls, storage and transfer of gasoline controls, petroleum storage controls	South Coast AQMD
Santa Barbara County APCD	New and modified source review, permit to construct, new source performance standards, stack monitoring, source sampling and testing, so controls, no controls, solvent and rvp controls, storage and transfer of gasoline controls, petroleum storage controls	Santa Barbara County APCD
Ventura County APCD	New and modified source review, permit to construct, new source performance standards, stack monitoring, source sampling and testing, so controls, no controls, solvent and rvp controls, storage and transfer of gasoline controls, petroleum storage controls	Ventura County APCD
San Luis Obispo County APCD	No data	San Luis Obispo County APCD

Table C.9
Laws and regulations pertaining to handling and storage of hazardous materials.
Source: A Profile of California's Oil and Gas Industry (California Department of
Conservation, Division of Oil & Gas, 1986).

	Requirement	Responsible agencies
Emergency Planning and Community Right-to-Know	Submittal of material safety data sheet, chemical inventory reporting, release reporting	US EPA, Local Emergency Planning Committee
US Code 49 CFR 394, Vehicle Code	Hazardous materials transport; accidental spills of waste of hazardous materials	Federal Department of Transportation, California Department of Emergency Services, California CHP
Occupational Safety and Health Act, California Labor Code	Personnel health and safety standards including training and records keeping	Federal and State OSHA
Hazardous Substances Storage Act	Regulations controlling the storage of hazardous substances including petroleum in aboveground storage tanks	California State Water Resource Control Board, Regional Water Quality Control Board
Federal Resource Conservation and Recovery Act, California Underground storage Tank Act	Regulations controlling the storage of hazardous substances including petroleum in underground storage tanks	US EPA, California State Water Resources Control Board, Regional Water Quality Control Boards, local agencies
California Health and Safety Code, Proposition 65	Regulations controlling the handling of hazardous materials including inventory reporting, and hazardous materials training	California OSHA, California Office of Emergency Services, California Sate Water Resources Board, Department of Toxic Substances Control, Office of Environmental Health Hazards Assessment

spill. The act established an oil spill prevention and response wing within the California Department of Fish and Game and funds it by taxing oil transported into California. The act also requires all coastal facilities to have an emergency spill response protocol in case of marine release. In addition to OSPRA, the California Division of Oil, Gas, and Geothermal Resources requires a spill contingency plan for most operations. These regulations require that tanks have control methods for spilled fluids, special safety devices for offshore and critical wells, and that operators maintain a database of spill incidents in order to stop spills from happening and improve spill response if and when they occur.

In the event of a spill, a report must be made to the California Office of Emergency Services, which then notifies the appropriate agencies. If a spill affects state waters, the Water Quality Control Boards are called. If the marine environment is affected, the Coast Guard is the lead agency. If a spill involves potentially hazardous materials, on land the state department of Toxic Substances is brought in. Several other agencies are involved in the case of a well blowout or discharge prevention, with Division of Oil, Gas, and Geothermal Resources being the lead in most instances. (See table C.9.)

Notes

Introduction

1. The Guadalupe Dunes have been designated a National Natural Landmark. This designation, conferred by the US Secretary of the Interior, acknowledges the national significance of the dunes as a exceptional and rare ecosystem.

2. I use 'spill' to refer to the total accumulation of diluent—the end product. I use 'spillage', 'leaks', and 'leakage' to denote the process by which diluent was chronically lost over time, eventually becoming an enormous spill.

3. For analyses of cases of community contamination and the struggle to seek legal redress, see Calhoun and Hiller 1988; Brown and Mikkelsen 1990; Hawkins 1983.

4. This is also very similar to what the medical establishment refers to as "postponement behavior." As used by medical practitioners, that term refers to patients' resistance to modifying their behavior(s) in the face of incrementally deteriorating health. (Examples include diabetics who will not restrict their eating habits and smokers who continue to smoke.)

5. Both Turner (1978, pp. 81–125) and Hewitt (1983, pp. 3–29) identify this proclivity to ignore what Turner refers to as the "incubation stage" of industrial crises.

6. Environmental assessment is variously addressed in the US through federal Environmental Impact Statements, state Environmental Impact Reports, federal and state Natural Resource Damage Assessments, corporation-sponsored Health and Environmental Risk Assessments, and, generally, risk assessment and cost-benefit analysis.

7. Giddens has fallen prey to a related critique of his own work. He has been faulted for excessive social-constructionist tendencies. Though he has attended to ecological crises as both a defining feature of "high modernity" as well as a significant problem that confronts us all, he does so almost exclusively from the vantage point of discourse. Giddens goes so far as to claim that nature has ceased to exist as a category external to human social life. The problem relates to the obvious contradiction that this implies: If nature no longer exists, why then are ecological "ills" a threat, aside from the social anxiety they may produce? It would seem, then, that from Giddens's perspective all we need do is convince ourselves that they are not occurring! (See Dickens 1996, p. 41.)

8. On the entrenchment of the "human exceptionalist paradigm" in the social science and in Western society in general, see Dunlap and Catton 1979 and Dunlap 2000.

9. On "invisible" contaminants and their ability to "traumatize," see Vyner 1988.

10. A Hollywood movie titled *The China Syndrome*, released 2 weeks before the event at Three Mile Island, played a part in popularizing nuclear issues and fears.

11. For instance, on the West Coast the Pacific Coast Federation of Fishermen's Association is a very powerful lobby. This group lobbies state legislatures from California to Alaska, as well as the US Congress, on the behalf of sport fishermen and commercial fishing interests. Another lobby, more local to Alaska, is Cordova District Fishermen United, an influential political player in Alaska resource policy.

12. I am as guilty of this as anyone. I did not become aware of the Guadalupe spill until the local news media reported that 500,000 gallons of petroleum had been recovered by Unocal.

13. For the national search, I used the University of California's MELVYL news database, which looks for words used in headlines. Five leading US newspapers are indexed: the *New York Times*, the *Los Angeles Times*, the *Washington Post*, the *Christian Science Monitor*, and the *Wall Street Journal*. I searched headlines for mentions of "*Exxon Valdez* Spill" and "Guadalupe Spill." Because the search turned up so few Guadalupe Spill articles, I then searched for "Unocal Oil Spills" and "California Oil Spills" (and various synonyms), finding three more stories.

14. Compare Mazur 1998.

15. Especially in a region where oil and environmentalism do not mix.

Chapter 1

1. Japan was importing a relatively large volume from the US around this time.

2. The military bases included Camp Roberts (the largest in the World at that time), Camp San Luis Obispo, Morro Bay Naval Base (east of San Luis Obispo), and Camp Cook (on the San Luis Obispo/Santa Barbara county line). See Krieger 1990.

3. The notoriously leaky piping system was installed in the early 1970s (Stormont 1956; Arthur D. Little et al. 1997).

4. In oil spills that involve criminal charges, chemical "fingerprinting" is routinely used to compare the petroleum discovered against that used or hauled by those suspected of being responsible. As in DNA testing, matching is based on the *probability* that the oil found matches that used or hauled by the suspect. Because different petroleum deposits carry signatures or "distinctive and patterned identifiers," comparisons and differentiation is possible. For instance, oil from California's central coast carries a different signature or fingerprint than does oil from Alaska's North Slope. However, these comparisons and the probabilities attached are not as precise as the term "fingerprinting" may imply. The scheme to compare and identify in court consists of three categories: "consistent with," "similar to," and "not consistent with." The only one that can sponsor a conviction is the first. This is especially problematic in cases where the oil is refined and its chemical composition is incon-

sistent and indefinite. Such is the case with petroleum spilled at Guadalupe (sources: email from toxicologist, Crocker Labs, University of California, 1998; Elliott 1999).

5. Threatened flora include the Marsh Sandwort, Gambells Atercress, La Graciosa Thistle, Surf Thistle, Beach Spectacle-Pod. Threatened wildlife, which are known to frequent the area, include the Steelhead Trout, Tidewater Goby, Peregrine Falcon, California Brown Pelican, California Least Turn, Western Snowy Plover. These are only those which are on the federal government's Endangered Species List. A list of some 40 other state and locally threatened plant an animal species are also of concern at the site (source: Arthur D. Little et al. 1997).

6. San Francisco and Long Beach are important markets for Surf-Perch (warden, California Department of Fish and Game, interviewed in 1997).

7. See also Stallings 1990; Gamson and Modigliani 1989.

8. When it did appear in local headlines, it was dubbed the "Silent Spill" (Ritea 1994).

9. Some 136,000 cubic yards of sand were removed at a depth of 20 feet and a total of 250,000 gallons of contaminant removed from this single site (Arthur D. Little et al. 1997).

10. Source: US District Court, Northern District of California, 1994.

11. Source: Interdepartmental memo, California Regional Water Quality Control Board, February 20, 1990.

12. "Daylighting" is a technical term that refers to forcing petroleum to the surface by means of fluctuations in the water table. Because diluent is lighter than water, it was forced to the surface of the dunes as the local water table surged with winter rains.

13. See also Jackowoy 1993.

14. An oil sump is a large depression or pit used to store oil contaminated sediments, drilling muds, and water pumped from extraction wells. Over the years a thick, gooey asphalt like layer of oil formed along the bottom of these sumps, which remain in and around the field (Sneed 1997).

15. To date, some 90 separate-phase (i.e., with petroleum floating free on top of the groundwater) petroleum plumes have been found at the site, with diluent in the dissolved phase (mixed with the groundwater and sand at the site) contaminating more than 60% of the 6-square-mile site (Sneed 1998; Arthur D. Little et al. 1997).

16. The federal acts that institutionalized self-reporting or self-monitoring and that are applicable to the Guadalupe spill are the following: section 311(b)(5) of the federal Water Pollution Control Act (i.e., the Clean Water Act), amended 1973; section 306(a) of the Outer Continental Shelf Lands Act, amended 1978; section 16(b) of the Deepwater Ports Act of 1974. All require the responsible party to notify the National Response Center in the event of an oil spill from a vessel or an operating facility. (These are primary statutes and do not represent all applicable state or federal level edicts that call on industry to police its own activities.) For a complete listing of regulations that address pollution generally and petroleum spills specifically, see appendix C. The following acts and codes were alleged by the State of California to have been violated by Unocal: Water Code subsections 13350(a)(2) and

13350(a)(3); Fish and Game Code sections 5650, 5650.1, 12015 12106, and 2014; Government Code sections 8670.25.4, 8670.66(a)(4), 88670.66(a)(3), 8670.56.6, and 12607.6; Harbors and Navigation Code section 151; Health and Safety Code sections 25249.5, 25249.6, 25189.2(c), 25189(d), 25189(b), and 25143.10; Business and Professions Code sections 17200; Civil Code section 3479, 3480, and 3481. In addition, there were claims of common-law natural-resource damages and negligence and claims under California's Proposition 65 "right to know" legislation (Superior and Municipal Courts of the State of California, County of San Luis Obispo. 1998, p. 2).

Chapter 2

1. For examples of theoretical work on the relationship between society and nature, see Dickens 1996; Freudenburg, Frickel, and Gramling 1995; MacNaghten and Urry 1995; Capra 1975, 1988; Merchant 1980; Schnaiberg 1980; Marcuse 1964; Mumford 1934.

2. There has also been much research on how science itself is conducted with social constituents in mind, exemplary are Collins and Pinch 1993; Latour and Woolgar 1986; Latour 1987; Longino 1990; Shrader-Frechette and McCoy 1993. There are also those that have followed technological innovation viewing it as an imbricated sociotechnical process (see also Bijker and Law 1997; Bijker, Hughes, and Pinch 1987; Hughes 1985).

3. See also Dunlap 2000; Woodgate 2000; Dickens 1996, p. 47; MacNaghten and Urry 1995; Szerszynski 1996.

4. For more of this type of research with its emphasis on communities and interpretation see Levine 1982; Szasz 1994; Edelstein 1988; Espland 1998; Kuletz 1998; Freudenburg and Gramling 1994.

5. Schnaiberg (1980) makes an important distinction between what he refers to as prestigious "production scientists" (real science, hard science, and precise science) and their association with industry, whose primary responsibility is the development of new applied "technologies" (technology in the broadest sense). This is in contradistinction with what he terms the "impact scientist," whose primary objectives are in ascertaining the impacts of technologies that have been introduced.

6. For an applied example, see the Ad Hoc Risk Assessment Review Group's 1978 evaluation of US Nuclear Regulatory Commission 1975. The evaluation was published by the NRC.

7. The most notorious example of this public protest and refutation of expert claims is that of nuclear power and the fights surrounding public fluoridation programs (Epstein 1991; Mazur 1981; Freudenburg and Pastor 1992; Martin 1991).

8. The waste management industry has functionally employed this model, aggressively pursuing locations with high unemployment and poor economies in the hope that promises of work and tax base support, coupled with strong claims of safety, will compensate and in so doing win local support for their facilities (Szasz 1996; Hofrichter 1993; Bullard 1990).

9. For a particularly clear example to compare against the position being taken here, see Starr 1969.

10. For research that documents community struggles for recognition and remuneration because of toxic contamination see Brown and Mikkelsen 1990; Calhoun and Hiller 1988; Harr 1995; Levine 1982. Questions concerning what is "proof" have a good deal of relevance for the Guadalupe spill. As touched on in the introduction, controversy surrounds whether diluent is a toxin, how toxic it is, and as an outgrowth what to do about it. Moreover, the damages (or lack thereof) that the spill is "assessed" as responsible for are at the heart of the legal case itself. In Guadalupe's case, as with toxic tort cases more generally (legal cases involving toxic contamination), because of the inexact makeup of different "batches" of diluent used at the field over the 38 years of production, it is virtually impossible to *precisely* identify (referred to as "fingerprinting,") and thus in an adversarial court room setting prove. Add to this fingerprinting problem the natural oil seeps which are in and around the area as well as other petroleum products which were spilled at the site by Unocal over the years and that are not part of litigation against Unocal for spilling diluent, and one gets a picture of how difficult it is to exactly identify "the problem."

11. Calculating the probability of a catastrophic outcome is, to someone who must live by that calculation, little reassurance. This is especially so if the source of the estimate is not a trusted one. I take up risk assessment, risk perception, and trust in institutions and authority, as it relates to community perceptions or risk, in chapter 5.

12. For example, in cases of contamination one must prove that they have been adversely affected in order to sue for damages. Yet, *toxic tort* cases, as they are referred to in legal circles, are notoriously difficult for the plaintiff to win, because cause and effect can be so very difficult to determine. See preceding note.

13. See Latour (1993) on "underdetermined science."

14. Resistance to litigation during the Reagan era, its persistence in the Clinton era (through Reagan's appointment of judges sympathetic to business interests and in the form of recent state legislation to limit both the number of suits and the size of settlements), and its re-emergence with the George W. Bush administration points out the difficulty the public confronts when seeking to participate, as well as seek redress, when their safety has been compromised through negligence or accident. Source: Derivative of a personal conversation with William Freudenburg, Department of Rural Sociology and Sociology, University of Wisconsin, Madison. See also McGinnis and Proctor 1999; Eckstein 1997; Clarke 1989.

15. See Lester Brown's *State of the World* series.

16. See, e.g., Kroll-Smith and Couch's (1990) empirical investigation of a creeping mine fire under a Pennsylvania town. These authors describe the circumstances that surround the mine fire's manifest qualities (i.e., its progressive destruction of the town); they also describe the fire's psychosocial impact on the community (i.e., they provide an analysis of subjective and interpretive processes). Their research highlights the nexus of environmental and social-contextual factors that are at play when communities confront extreme and toxic environments.

17. See also Walsh 1988, 1981; Erikson 1994; Edelstein 1993, 1988; Freudenburg 1991.

18. In the Japanese movie *Rashomon*, multiple perspectives on the same event—a crime—are described in a court. Each recounting is the product of the speaker's standpoint.

19. On this point of rational vs. irrational, while Mazur (1998) does not equate irrationality with risk interpretations, he does seek to establish which are "truer" than others.

20. For example, this is the form the cost-versus-benefits analysis forwarded by the Nuclear Regulatory Commission in the late 1970s concerning nuclear power generation (see Cohen 1985).

21. See also Wynne 1996, 1992.

22. There is an important and glaring exception to a sympathetic social-constructionist position on laypersons risk interpretations. Mary Douglas and Aaron Wildavsky (1995), in *Risk and Culture*, relied on a social-constructionist argument to point out the irrationality of modern environmental fears. Because such fears, according to Douglas and Wildavsky, are the outgrowth of a modern and thriving "culture of distrust" (in authority) and not systematic appraisal, they concluded that such fears were largely irrational assessments of danger—nothing less than modern magic. They likened such belief systems (through anecdote) to primitive beliefs and superstitions. For a range of perspectives on the social construction of risk see Dickens 1996; Hilgartner 1992; Rayner and Cantor 1987; Rayner 1987; Clarke 1989; Gamson and Modigliaini 1989; Stallings 1990; Freudenburg and Jones 1991; Short and Clarke 1992.

23. See also Couch and Freudenburg 1999.

24. For a traditional disaster research recap, see Fisher 1998. See also Hewitt 1983; Fritz 1961; Quarantelli and Dynes 1972, 1977.

25. See Quarantelli 1978; Quarantelli and Dynes 1972, 1977; Couch and Kroll-Smith 1991.

26. It is quite easy to understand why, in the case of natural disasters, this model works well as it closely resembles the operational definition employed by disaster researchers: people have little control over a disasters arrival, occurrence, or departure.

27. As has already been noted, these chemicals are in the diluent spilled at Guadalupe.

28. See also Clarke 1999.

29. That is, "risky" (Vaughan 1996, p. 408).

30. See Sagan 1993 and Snook 2000.

31. See La Porte and Consolini 1991; Weick and Roberts 1993; Weick 1987; Roberts 1982; Wildavsky 1988; Morone and Woodhouse 1986.

32. 'Hazard', as defined by La Porte and Consolini (1991, p. 23), "refers to the characteristics of a production technology such that it fails significantly the damage to life and property can be very considerable."

Chapter 3

1. For more details, see appendix C.

2. A barrel of oil is equivalent to 42 gallons of petroleum.

3. See also Perrow 1991.

4. For more on the "amoral calculator" thesis, see Kagan and Scholz 1984; Vaughan 1998, 1996; Lee and Ermann 1999; Calhoun and Hiller 1988; Szasz 1994.

5. The prices I use are those calculated for South Coast crude oil in the 1980s. Crude oil would be worth less than diluent, diesel, kerosene, or gasoline all of which must be refined in order to be used as a thinners or fuels. See Nevarez et al. 1996.

6. The Avila Beach Tank Farm spill is another Unocal problem in San Luis Obispo County. See Unocal 1997; Cone 1998.

7. Complete removal of the site's contamination is not possible. Over the first 10 years, the remediation plan calls for the excavation of a select number of problem sites—this is the "first phase" of cleanup. The second phase—10 more years—will involve pilot testing a number of remedial strategies and choosing the one deemed most effective. In the end, nature and time will be the only remedies to the extensive contamination of the site.

8. In 1996, Unocal set aside $250 million to handle environmental remediation worldwide (Unocal Corporation 1997).

9. See also Aldrich (1999, p. 141) on communities of practice.

10. My interest is to better understand how more than 40 individuals who worked the field did not recognize the leaks as a problem and once they did why they kept quiet. Even if one can assume the six managers (superintendent, supervisors, and foremen) who ran the field were bad characters, I do not think the same assumption can be made of the entire fields workforce. The question "How and why do good people do dirty work?" (originally asserted so eloquently in Vaughan 1996) is the impetus for this chapter.

11. See King and Lenox 2000 for a recent article that follows compliance in the petroleum and chemical industries over time. In agreement with the findings of this chapter, King and Lenox have found, through industry-wide quantitative accounting, that leaving compliance to self-report does not sponsor cooperation and appears to be associated with the a reverse trend: higher rates of non-compliance.

12. See Weber 1952. See also Merton 1936; Perrow 1986.

13. See also Molotch 1990.

14. Historically oil work was a semi-autonomous activity left to the skills of a head driller and his hand picked crew of workers at each field (see also Romo 2000; Quam-Wickham 1994; Yergin 1991; Davidson 1988; Haslam 1972; White 1962).

15. For the sake of comparison, a circa 1900 perspective on oil in the Gulf of Mexico and early California is informative. Oil "gushers" evoked images of progress, events that we now refer to as oil spills. The historian Joseph Pratt (1978, p. 4) writes of this period: "In the rush for instant wealth, oil wealth, oil was seen

as black gold, not black sludge. The gusher was an ecological symbol for this early period—photographs of the Lucas Gusher went out across the world, showing the magnificent spectacle of the 6-inch stream of oil rising more that 100 feet above the top of the derrick. So powerful was the image that wise well owners arranged to turn on similar gushers for the entertainment and persuasion of potential investors. In this atmosphere of uncontrolled exploitation, few cameras recorded what happened to the gushing oil after it splashed to earth."

16. See also Clarke and Perrow 1997.

17. See also Pentland and Rueter 1994.

18. The vestigial clumps of crude oil are still observable in areas throughout the oil field. In 1997 I asked a Unocal chaperon during one of the site tours where the asphalt clumps had come from. He replied that Unocal's practice until the 1970s had been to spray the field's dunes with crude oil, especially along roadways, to keep them from shifting.

19. My visits to oil fields, my interviews concerning other fields in the region, and historical accounts of oil work all indicate that petroleum production has been a messy business since its origins in the late nineteenth century. See also Pratt 1978 1980; Dinno 1999.

20. There were also seniority-related grades within these five levels (field worker, telephone interview, 1997).

21. See Rice 1994; Paddock 1994a; Greene 1993a,b.

22. On "coupling" see Weick 1976, 1995; Perrow 1984.

23. On "whistleblowing" see Glazer 1987; Bensman and Gerver 1963; Elliston et al. 1985.

24. I encountered similar resistance to being interviewed on the part of oil-field personnel. One of the major constraints involved legal liability. Most potential respondents who represented the legal side (state Attorney General, Unocal lawyers) as well as both current and former Unocal employees were largely unwilling to speak to me, formally at least, about the spill because of potential liabilities.

25. The average yearly income in California's petroleum industry was $40,000 in 1990—this for a job that requires no college education. In comparison, the average yearly income for comparable jobs such as retail workers in California was $13,000 and for the manufacturing sector $25,000 (California Statistical Abstracts; Romo 2000).

26. On social, cultural, organizational inertia, see Vaughan 1996; Becker 1995; Hannan and Freeman 1984.

27. This is also in line with Bensman and Gerver's (1963) articulations of "institutional schizophrenia."

28. See Garfinkel 1956.

29. The first whistleblower was fired a few months after reporting the spill and could find only intermittent work in the industry thereafter.

30. In the end, all of this occurred.

31. On environmental law in the US, see Yeager 1993; Wolf 1988; Skillern 1981. When taken to court in 1992 and 1993, Unocal did not face a single charge, but 28 separate criminal and misdemeanor violations.

32. For more on deviance and solidarity in groups and organizations, see Sherman 1987; Katz 1977; Becker 1963; Gouldner 1954; Dentler and Erikson 1959.

33. See also Collinson 1999; Lee and Ermann 1999; Clarke 1999.

Chapter 4

1. Though this is difficult to verify precisely, it seems that the anonymous field worker and a beach walker (or walkers) called the California Office of Emergency Services, who in turn called the Coast Guard about possible ocean releases, who in turn informed the California Department of Fish and Game.

2. I focus on the federal level because most state environmental legislation has followed on the heels of federal edicts. Furthermore, federal agencies have been primary in setting environmental compliance standards for the states to enforce, even if states such as California have eschewed in certain instances those standards in favor of more rigorous programs (Turner and Rylander 1997; Colella 1981). According to Colella (1981), the federal government established itself as the "environmental protector" through federally enforceable legislation during the 1960s, and in so doing provided for a baseline in environmental compliance.

3. Especially with the Clean Water Act amendments of 1970, 1972, and 1977.

4. The Environmental Protection Agency is the primary example, but within agencies, reorganizations took place to address such newly highlighted environmental concerns.

5. On the oil industry's response to such legislation, see Beamish et al. 1998; Yergin 1991.

6. See EPA 1974, p. ix.

7. OPA also instituted tougher penalties, outlined responsible party liabilities, and allocated more resources to spill response systems.

8. The other assumption being that anything more than 10,000 gallons would be obvious and receive emergency response from the appropriate resource or government agency. In view of Guadalupe's history, this is also a faulty assumption.

9. The California State Division of Oil, Gas, and Geothermal Resources rely on self-reporting even more than the other agencies.

10. See Granovetter 1985; Perrow 1986; Scott 1981; Thompson 1967. On "implementation" in organizational contexts and that problematizes prescriptive theories of "rational decision making," see Diver 1980; Schulman 1989; Ermann and Lundman 1978; Wildavsky and Pressman 1984.

11. "Technology," as conceptually used by organizational theorists, is not isolated to hardware such as equipment, machines, and instruments. Rather the concept involves a broader view that includes codified skills and knowledge such as

organizational routines and institutionalized modes of conduct (Scott 1998; Perrow 1986; Nelson and Winter 1982).

12. See also March and Olsen 1979.

13. See also Perrow 1986.

14. Clarke (1989, p. 174) also "doubt[s] that the elements of an organizational system are as randomly organized as garbage can theory would have us believe." This also holds for response to the Guadalupe spill, where the problem, the solutions, and the choices made did not appear "random" per se, even if they were not optimal or closely aligned with what could or even should have been done.

15. OPA 1990 is a significant improvement over earlier emergency spill-response protocols, even if the act's prescriptions did not match the scenario that unfolded at the Guadalupe Dunes (US Coast Guard 1995a).

16. Source: San Luis Obispo District Court, case M-202983, December 20, 1993.

17. In 1990, regulators still did not know that the field was extensively contaminated.

18. A bioassay is a technical assessment of the biological integrity of an area.

19. The unit went on line in 1990 as a consequence of the Lempert-Keene-Seastrand Oil Prevention and Response Act, which was enacted in 1990 to complement the federal Oil Pollution Act of 1990 (OPA). The Office of Oil Spill Prevention and Response (OSPR) is a California State Department of Fish and Game task force that specifically targets emergency petroleum releases.

20. Highway 1, which runs along the Pacific Coast, at this location is approximately 3 miles from the spill site and the ocean.

21. At this time, the Coast Guard saw the field's contamination as a set of land based leaks. They will soon change their position on this, reinterpret the leakage as a marine spill, and for a short time "take charge."

22. See US Coast Guard 1995a.

23. Ocean resources were legislatively protected; the dunes were not.

24. Traces of diluent have been found as far as 20 miles from the oil field in ocean waters routinely tested by regional offshore oil platforms, showing how chronic, extensive, and unmitigated leakage has been (public presentation, lead consultant, Arrthur D. Little Inc.).

25. See also Bunin 1994. This surprised the Coast Guard contingent, the lead agency for the 1993–94 beach excavation project. Moreover, it is interesting to note how little the Coast Guard learned from this public outcry and the lucid criticism voiced concerning impacts to the ecology of the site. In a follow-up assessment of their response, the Coast Guard generally congratulated themselves with one small admonition, they had not "fully anticipated the intensity of the public perception controversy that grew up around their response. Public affairs must be viewed as a critical success factor and must be employed proactively. . . ." (US Coast Guard 1995b) The implication of these comments, according to Coast Guard Officials, is that such public relations tactics must be employed proactively to avoid misunderstandings on the part of an uninformed public.

26. In the case of the *Torrey Canyon*, British authorities had little time, so they applied the solution they had for a problem they knew little about and inflicted damages equally as great as the spill itself. In the case of the *Exxon Valdez*, the PR debacle forced actions based on images of ruined beach's rather than on deeper ecological issues concerning the long-term implications of quick fixes. For a sobering assessment of oil spill response, in particular those that followed the *Exxon Valdez* accident, see Clarke 1990.

27. This is not to say that the Coast Guard was wrong in any objective sense. There is no current consensus on the impacts of the chronic leakage of diluent into the dunes and the ocean.

28. State and local regulators have forced Unocal to comply with California's Environmental Quality Act (CEQA) to assure that the least damaging environmental cleanup method is chosen, while assuring that the most thorough remediation is accomplished. This is only the sixth time that CEQA has been utilized to assess the cleanup strategies proposed by those responsible for a pollution event. Generally, an Environmental Impact Report (EIR) is required by CEQA when a project has the potential for significant environmental impact. CEQA applies to any activity proposed, funded, or permitted by a state or local agency that has the potential for resulting in physical change in the environment. The CEQA process, as it is referred to, normally consists of three parts or phases. The first phase consists of a preliminary review of a project to determine whether it is subject to CEQA. The second involves preparation of an initial study to determine whether the project may have a significant environmental effect (if not, a negative declaration). The third phase is the preparation of an EIR if the project is determined to have significant effects. (See appendix C for details.)

29. For a clear articulation of this tendency on the part of a range of institutional bodies, see Clarke 1999.

Chapter 5

1. Here I focus on the south and southwest of the county, which is socially, economically, demographically, and thus culturally distinct from the northern and northeastern parts (Beamish et al. 1998; Nevarez et al. 1996).

2. In another example, a retiree and Guadalupe/Nipomo Dunes advocate related his trepidation in returning to his boyhood home back East, "now that he was aware of the pollution that is rife there." In interviews, impromptu conversations, and ethnographic contexts, San Luis Obispans consistently echoed similar sentiments, that it was "not anywhere near as bad here" (as in other places). A feeling that motivated their very defensive attitude with regard to their Central Coast home and its natural environment: "I used to live in Pittsburgh, Pennsylvania. It is a long-term industrial town. . . . Steel is a dirty industry. Mining is a dirty industry. You deal with a lot of toxic materials. . . . I moved back there in 1978, spent 7 years back there working. . . . In going back, we were concerned about water supply. In the East in general, most cities get their water supplies from the rivers. . . . You have all kinds of industries back there that dump toxics illegally all over the place. It was expedient.

It was cheap. I have seen rivers run crystal clear because mine tailings had poisoned the river. You couldn't get algae or anything to run in that river. It was just like distilled water. It is a little scary; we have to be cautious about that. I don't think it is anywhere near as bad here as it is in other places." (resident of San Luis Obispo County, interviewed in 1997)

3. For instance, as discussed in the opening other well publicized cases of community's and toxic contamination include those of Woburn, Massachusetts; Love Canal, New York; and Times Beach, Missouri.

4. Over the years there have been developments slated for the dunes. Two examples are the proposal to build a nuclear plant there (it was later installed at Diablo Canyon, I discuss this in subsequent pages) and, in the early 1980s, Husky Oil's (another, but small-time producer at the dunes) proposed expansion of their oil operations.

5. For excellent reviews of the risk literature, see Clarke and Short 1993; Freudenburg and Pastor 1992; Shrader-Frechette 1991.

6. "Oil Spill Estimates Buried by Reagan" (anonymous editorial, *Telegram-Tribune*, April 7, 1989).

7. See Solen 1998 for details on Santa Barbara's attempts to forestall petroleum development.

8. The industry outspent the proponents of Measure A tenfold.

9. Reed and Foran (1999) and Foran (1997) use "political cultures of opposition" to better understand the nascent roots to revolutionary movements.

10. For the period 1989–1996, I found 325 stories that focused on "petroleum" as a topic area. Eight stories covered *oil* as a relatively neutral and specific topic area, 246 within *oil spills* as a specific topic area, and 71 under *offshore oil* as a specific topic area. Because the *Telegram-Tribune*'s story index is organized differently before and after 1991, the method I used to search for and locate stories varied slightly. For the years 1989 and 1990, I searched the *Telegram-Tribune* story index for "oil stories" using the following keywords (in alphabetical order): *local environmental issues, hazardous materials, offshore oil, oil, oil drilling, oil tanker(s), petroleum.* Most of the stories found for these first 2 years were in the subcategory *local environmental issues.* Because stories within the index are cross-referenced and can fall within multiple categories, my search included scanning each story's headline to ensure that each was counted only once. (Each story is indexed under a subsection title, then by headline, then by section and page number, and then by column number. For instance, stories concerning Guadalupe's spillage could be found under the subtitles *Guadalupe* and *Guadalupe Dunes* as well as under *oil spills.*) Beginning in 1991 and continuing through 1996, the categories *oil, oil spills,* and *offshore oil* were consistent across years and continued to be cross-referenced with other index categories in which such stories could fall. After 1991, because I found the categories *oil, oil spills,* and *offshore oil* to be inclusive, I tabulated the number of stories found in these and no other categories.

11. Although primarily directed at corporate and governmental institutions, county distrust of outside intentions is not exclusive to them. In 1998, the Los Angeles based

legal council of the Surfrider Foundation initiated a suit directed at Unocal (as well as state, and—very important—local regulators) for polluting coastal waters and not fulfilling the public trust in the name of local surfers. Some locals have bristled. For example: "I'm on the side of Unocal as regards Guadalupe Dunes, surprised? That surfing group appears to be out for money and nothing else. Their claims are ridiculous!" (email from resident of San Luis Obispo County, 1998) Though this respondent's statement was unusual, it does reveal a deeper sense of who is to be trusted that was mirrored in the comments of others with whom I have spoken.

12. The investigation's initial failure in court, according to those involved—including the California Fish and Game wardens themselves—rested on a Fish and Game investigation that was slow to manifest. I have taken up why in the introduction and the previous chapter, so suffice it to say the wardens relied on self-report to an extent that it inhibited their ability to freely investigate the spill. This culminated in the dismissal of charges. County residents do not necessarily see it in this light as this chapter makes the case.

13. For research that speaks directly to this sense of "victimization" see Edelstein 1993; Couch and Kroll-Smith 1985; Kroll-Smith and Couch 1993.

14. For explicit treatment of this relationship between risk and trust in modernity, see Beamish 2001; Lash et al. 1996; Wynne 1996, 1992, 1987; Beck et. al. 1994; Erikson 1991; Giddens 1990, 1991

15. For critical research on risk evaluation methods, see Shrader-Frechette 1997 1995; Finkel 1996; Freudenburg 1994 1988; Clarke 1993b; Rayner and Cantor 1987.

16. For an excellent discussion, see Freudenburg 1993.

17. For instance, those posed by nuclear technologies (see Starr 1969).

Chapter 6

1. See Camic 1986 (on Durkheim, p. 1056; on Weber, p. 1066).

2. Bourdieu is particularly interested in social differentiation and the stability of social class as a primary axis for "distinction." His development and distinction of different types of capital—social, economic, and cultural—lie at the heart of his theoretical project. See Bourdieu 1984.

3. See Latour 1987; and other science studies scholars: Law 1992, 1987; Bijker and Law 1997; Hilgartner 1992. Molotch, Freudenburg, Paulsen (2000) offer a variant on this system/network agglomeration model describing a weaving (Law 1987) or lashing together (Latour 1987) of cultural attributes and "place" to explain the divergent growth trajectories of two Southern California cities with virtually identical physical-topographic characteristics. In a similar vein, recent theorists of economy and society have also evoked terms that, if they do not name habit precisely, rely on synonyms such as convention, routines, tradition, and "indigenous institutional arrangements" to express how such conventionally maligned social elements cut both ways (Biggart 1999; Biggart and Guillen 1999; Storper and Salia 1997; Saxenian 1994).

4. See also Stinchcombe 1965, p. 148; Hannan and Freeman 1984, p. 160.

5. In 1986 and 1988 local Unocal managers denied that the petroleum release had originated at their field.

6. A local activist summed up local sentiment as follows: "Unocal has always been a bad actor: spills along its pipelines in the county . . . refusing to clean them up. There was no love lost to begin with. . . . It was more like 'These guys are screwing us again! They're polluting: they're lying. They don't want to clean it up. They have been hiding it for so many years, and it's unacceptable!' And that's when we started . . . digging into it . . . getting people's outrage together."

7. This recalls Walton's (1992b) articulation of "traditions of resistance" and Reed and Foran's (1999) "political cultures of opposition."

8. With roots in Rousseauian social philosophy the concept of a social charter has been developed by labor movement theorists such as Reinarmen (1987), Flacks (1988), and Fantasia (1988) to capture a bargain struck between the state, big business, and civil society. This notion of a violated social charter also resonates with research on human-induced crisis (Freudenburg 1993; Erikson 1976 1994; Wynne 1992; Kroll-Smith and Couch 1990; Edelstein 1988; Brown 1992; Brown and Mikkelsen 1990. See also Granovetter 1985; Shapiro 1987 for reference to trust in exchange relationships that are economic is character).

9. Examples include urban and agricultural pesticide runoff, multi-source CFC-generated ozone depletion, and CO_2-induced climate change.

Appendix A

1. According to Gould (1986, p. 64), "the great philosopher of science William Whewell had called this historical method 'consilience of induction.'"

2. I used a text-based database manager—FolioViews—to organize the interviews, conduct string searches, and save conceptual thematics. The program, like a burgeoning list of like software applications, allowed for quick retrieval, quick searches, and the ability to organize data with cross-references (a.k.a. jump links) and other useful coding procedures. The program allows for qualitative intuition to "run the show," but gave the speed and thoroughness associated with more quantitative procedures often associated with programs such as SASS.

3. I gained direct interview access to four past and present workers at the Guadalupe field. To verify and better ensure the reliability of their comments, I cross-referenced assertions with other interviewees (including those of regulatory officials who had interviewed field workers during their investigations of the spill), with other interviews with Unocal workers who worked at other locations, with press accounts of workers' impressions, and with historical and contemporaneous accounts of oil work. In addition to these four former Guadalupe field hands, I had access to three transcripts of interviews with oil-field workers from the region who had worked with or for Unocal (although not at the Guadalupe field) that included comments relevant to my analysis. Also relevant to my analysis, outside of research on the Guadalupe spill, is the fact that from 1990 through 1998 I participated in a

research project that focused on the Central and Southern California oil industry and on local community development. In this research endeavor, I interviewed regional persons involved in all aspects of petroleum production and government oversight; in doing so, I gained a thorough understanding of petroleum production as an industrial endeavor.

4. This was one of the few opportunities I had to talk with the legal staff, outside of an interview with a trial lawyer representing local surfers suing Unocal.

5. My attempts to obtain video tapes from the three local TV stations, or at least scripts used to depict spill events, came to naught; the prices were too high. However, in view of the cross-cutting nature of newspaper and television production (Gans 1980; Entman 1989; Fishman 1978), the major frames of reference were assumed to be minimal.

Appendix C

1. Proposition 65, better known as the "right-to-know act," holds industrial polluters legally accountable for failing to report effluent release into the environment to surrounding and potentially affected communities.

References

Aldrich, H. 1999. *Organizations Evolving*. Sage.

Arthur D. Little Inc. in association with Furgro West, Headley and Associates, Marine Research Specialists, and Science Applications International Corp. 1997. Guadalupe Oilfield Remediation and Abandonment Project.

Asch, S. 1951. "Effects of Group Pressure upon the Modification and Distortion of Judgments." In *Groups, Leadership, and Men*, ed. H. Guetzkow. Carnegie Press.

Bauman, Z. 1992. "The Solution as Problem." *Times Higher Education Supplement*, November 13: 25.

Beamish, T. 2000. "Accumulating Trouble: Complex Organizations, a Culture of Silence, and a Secret Spill." *Social Problems* 47, no. 4: 473–498.

Beamish, T. 2001. "Environmental Hazard and Institutional Betrayal: Lay-Public Perceptions of Risk in the San Luis Obispo County Oil Spill." *Organization and Environment* 14, no. 1: 5–33.

Beamish, T., H. Molotch, P. Shapiro, and R. Bergstrom. 1998. Petroleum Extraction in San Luis Obispo County: An Industrial History. Final Report, Project 14-35-0001-30796, Department of Interior, Minerals Management Service, Pacific OCS Region. Ocean Coastal Policy Center, Marine Science Institute, University of California, Santa Barbara.

Beck, U. 1992a. *Risk Society: Towards a New Modernity*. Sage.

Beck, U. 1992b. "From Industrial Society to the Risk Society: Questions of Survival, Social Structure and Ecological Enlightenment." *Theory, Culture & Society* 9: 97–123.

Beck, U. 1996. "Risk Society and the Provident State." In *Risk, Environment and Modernity*, ed. S. Lash et al. Sage.

Beck, U., A. Giddens, and S. Lash. 1994. *Reflexive Modernization: Politics, Tradition, and Aesthetics in the Modern Social Order*. Polity.

Becker, H. 1963. *Outsiders: Studies in the Sociology of Deviance*. Free Press.

Becker, H. 1995. "The Power of Inertia." *Qualitative Sociology* 18: 301–309.

Bensman, J., and I. Gerver. 1963. "Crime and Punishment in the Factory: The Function of Deviancy in Maintaining the Social System." *American Journal of Sociology* 28: 588–598.

Berger, P., and T. Luckmann. 1966. *The Social Construction of Reality: A Treatise in the Sociology of Knowledge*. Doubleday.

Biggart, N., and M. Guillen. 1999. "Developing Difference: Social Organization and the Rise of the Auto Industries or South Korea, Taiwan, Spain, and Argentina." *American Sociological Review* 64: 722–747.

Bijker, W., and J. Law, eds. 1997. *Shaping Technology/Building Society: Studies in Sociotechnical Change*. MIT Press.

Bijker, W., T. Hughes, and T. Pinch, eds. 1987. *The Social Construction of Technological Systems*. MIT Press.

Birkland, T. 1998. "In the Wake of the *Exxon Valdez*: How Environmental Disasters Influence Policy." *Environment* 40, no. 7: 4–9, 27–32.

Bondy, C. 1994. "Guadalupe Isn't the First." *San Luis Obispo New Times*, February 23.

Boorstin, D. 1985. *The Discoverers: A History of Man's Search to Know His World and Himself*. Vintage Books.

Bourdieu, P. 1977. *Outline of a Theory of Practice*. Cambridge University Press.

Bourdieu, P. 1984. *Distinction: A Social Critique of the Judgement of Taste*. Harvard University Press.

Bradely, R. 1996. *Oil, Gas, and Government*. Rowman and Littlefield.

Brown, L. 1999. *State of the World: A Worldwatch Institute Report on Progress toward a Sustainable Society*. Norton.

Brown, M. 1980. *Laying Waste: The Poisoning of America by Toxic Chemicals*. Pantheon.

Brown, P. 1992. "Popular Epidemiology and Toxic Waste Contamination: Lay and Professional Ways of Knowing." *Journal of Health and Social Behavior* 33, September: 267–281.

Brown, P., and E. Mikkelsen. 1990. *No Safe Place: Toxic Waste, Leukemia, and Community Action*. University of California Press.

Brulle, R. 1995. "Environmentalism and Human Emancipation." In *Social Movements: Critiques, Concepts, and Case Studies*, ed. S. Lyman. New York University Press.

Bullard, R. 1990. *Dumping in Dixie: Race, Class, and Environmental Quality*. Westview.

Burger, J. 1997. *Oil Spills*. Rutgers University Press.

Buttel, F. 1987. "New Directions for Environmental Sociology." *Annual Review of Sociology* 13: 465–488.

Cable, S., and E. Walsh. 1991. "The Emergence of Environmental Protest: Yellow Creek and TMI Compared." In *Communities at Risk*, ed. S. Couch and S. Kroll-Smith. Lang.

Calhoun, G., and H. Hiller. 1988. "Coping with Insidious Injuries: The Case of Johns-Manville Corporation and Asbestos Exposure." *Social Problems* 35, no. 2: 162–181.

California Coastal Commission. 1999. Guadalupe Oil Field Spill Remediation Project Status Report dated "April 13, Th 11b."

California Regional Water Quality Control Board. 1988. Spill Report. File 1-005. Filed by Gregory W. Moon, Regional Environmental Affairs Manager, Unocal Oil Company. Submitted to California Regional Water Quality Control Board February 11.

Camic, C. 1986. "The Matter of Habit." *American Journal of Sociology* 91, no. 5: 1039–1087.

Capra, F. 1975. *The Tao of Physics*. Shambhala.

Capra, F. 1988. *The Turning Point: Science, Society, and the Rising Culture*. Bantam Books.

Carson, R. 1962. *Silent Spring*. Houghton Mifflin.

Chertow, M., and D. Esty. 1997. *Thinking Ecologically*. Yale University Press.

Church, T., and R. Nakamura. 1993. *Cleaning Up the Mess: Implementation Strategies in Superfund*. Brookings Institution.

Clarke, L. 1989. *Acceptable Risk? Making Decisions in a Toxic Environment*. University of California Press.

Clarke, L. 1990. "Oil Spill Fantasies." *Atlantic Monthly*, November: 65–77.

Clarke, L. 1993a. "The Disqualification Heuristic: When Do Organizations Misperceive Risks?" *Research in Social Problems and Public Policy* 5: 289–312.

Clarke, L. 1993b. "Context Dependency and Risk Decision Making." In *Organizations, Uncertainty, and Risk*, ed. J. Short and L. Clarke. Westview.

Clarke, L. 1999. *Mission Improbable: Using Fantasy Documents to Tame Disaster*. University of Chicago Press.

Clarke, L., and W. Freudenburg. 1993. "Rhetoric, Reform, and Risk." *Society*, July-August: 78–81.

Clarke, L., and J. Short. 1993. "Social Organization and Risk: Some Current Controversies." *Annual Review of Sociology* 19: 375–399.

Cohen, B. 1985. "Criteria for Technology Acceptability." *Risk Analysis* 5: 1–3.

Cohen, M., J. March, and J. Olsen. 1972. "A Garbage Can Model of Organizational Choice." *Administrative Science Quarterly* 17: 1–25.

Colella, C. 1981. "Protecting the Environment: Politics, Pollution, and Federal Policy." In *The Federal Role in the Federal System*, ed. T. Conlan et al. Advisory Commission on Intergovernmental Relations.

Collins, H., and T. Pinch. 1993. *The Golem: What Everyone Should Know About Science*. Cambridge University Press.

Collinson, D. 1999. "'Surviving the Rigs': Safety and Surveillance on North Sea Oil Installations." *Organizational Studies* 20, no. 4: 579–600

Cone, M. 1998. "Unocal to Pay $43.8 Million Fine in Spill." *Los Angeles Times*, July 22.

Connell, S. 1999. "Huge Oil Spill Cleanup Advances." *Los Angeles Times*, November 22.

Couch, S., and W. Freudenburg. 1999. "Dancing, Diving, or Dealing with the Environment: A Devil or a Deep Blue Sea?" *Environment, Technology, and Society Newsletter*, spring: 3–6.

Couch, S., and, and S. Kroll-Smith. 1985. "The Chronic Technical Disaster: Toward a Social Scientific Perspective." *Social Science Quarterly* 66: 564–575.

Couch, S., and S. Kroll-Smith, eds. 1991. *Communities at Risk: Collective Responses to Technological Hazards*. Lang.

Couch, S., and S. Kroll-Smith. 1994. "Environmental Controversies, Interactional Resources, and Rural Communities: Sitting versus Exposure Disputes." *Rural Sociology* 59: 25–44.

Crenson, M. 1971. *The Un-Politics of Air Pollution: A Study of Non-Decision Making in the Cities*. Johns Hopkins University Press.

Davidson, R. 1988. *Challenging the Giants*. Oil, Chemical, and Atomic Workers International Union.

Davis, M. 1989. "Explaining Wrongdoing." *Journal of Social Philosophy* 20: 74–90.

Decarli, J. 1994. "Just When You Thought It Was Safe." *San Luis Obispo New Times*, April 13.

De Maria, W., and Cyrelle, J. 1996. "Behold the Shut-Eyed Sentry! Whistleblower Perspectives on Government Failure to Correct Wrongdoing." *Crime, Law and Social Change* 24: 151–166.

Dentler, A., and K. Erikson. 1959. "The Function of Deviance in Groups." *Social Problems* 7, no. 2: 98–107.

Dickens, P. 1996. *Reconstructing Nature: Alienation, Emancipation, and the Division of Labor*. Routledge.

Dinno, R. 1999. Protecting California's Drinking Water from Inland Oil Spills. Planning and Conservation League, Sacramento.

Dittmar, H. 1992. *The Social Psychology of Material Possessions: To Have It Is to Be*. Harvester Wheatsheaf/St. Martin's.

Diver, C. 1980. "Theory of Regulatory Enforcement." *Public Policy* 28, no. 3: 259–299.

Douglas, M., and A. Wildavsky. 1982. *Risk and Culture*. University of California Press.

Dunlap, R. 2000. "The Evolution of Environmental Sociology: A Brief History and Assessment of the American Experience." In *The International Handbook of Environmental Sociology*, ed. M. Redclift and G. Woodgate. Elgar.

Dunlap, R., and W. Catton. 1979. "Environmental Sociology." *Annual Review of Sociology* 5: 243–273.

Durkheim, E. 1984. *The Division of Labor in Society*. Free Press.

Easton, R. 1972. *Black Tide: The Santa Barbara Oil Spill and Its Consequences*. Delacorte.

Eckstein, R. 1997. *Nuclear Power and Social Power*. Temple University Press.

Edelstein, M. 1988. *Contaminated Communities: The Social and Psychological Impacts of Toxic Exposure*. Westview.

Edelstein, M. 1993. "When the Honeymoon is Over: Environmental Stigma and Distrust in the Siting of a Hazardous Waste Disposal Facility in Niagara Fall, New York." *Research in Social Problems and Public Policy* 5: 75–96.

Eder, K. 1990. "The Rise of Counter-Culture Movements Against Modernity: Nature as a New Field for Class Struggle." *Theory, Culture and Society* 17: 21–47.

Einstein, D. 1994. "Unocal, Officials Square Off on Spill." *San Francisco Chronicle*, March 15.

Elliott, G. 1999. "Shunning the Tarbaby." *California Coast and Ocean* 15, no. 3: 24–31.

Elliston, F., J. Keenan, P. Lockhart, and J. Van Schaick. 1985. *Whistleblowing: Managing Dissent in the Workplace*. Praeger.

Enloe, C. 1975. *The Politics of Pollution in a Comparative Perspective: Ecology and Power in the Four Nations*. David McKay.

Entman, R. 1989. *Democracy Without Citizens*. Oxford University Press.

Environmental Protection Agency. 1974. Disposal of Hazardous Wastes (Report to Congress pursuant to Section 212 of the Solid Waste Disposal Act, as amended). Government Printing Office.

Epstein, B. 1991. *Political Protest and Cultural Revolution: Nonviolent Direct Action in the 1970s and 1980s*. University of California Press.

Erikson, K. 1976. *Everything in Its Path: Destruction of Community in the Buffalo Creek Flood*. Simon & Schuster.

Erikson, K. 1990. "Toxic Reckoning: Business Faces a New Kind of Fear." *Harvard Business Review* 90: 118–126.

Erikson, K. 1991. "A New Species of Trouble." In *Communities at Risk*, ed. S. Couch and S. Kroll-Smith. Lang.

Erikson, K. 1994. *A New Species of Trouble: The Human Experience of Modern Disasters*. Norton.

Ermann, D., and R. Lundman, eds. *Corporate and Governmental Deviance: Problems of Organizational Behavior in Contemporary Society*. Third edition. Oxford University Press.

Espland, W. 1998. *The Struggle For Water: Politics, Rationality, and Identity in the American Southwest*. University of Chicago Press.

Etzioni, A. 1967. "Mixed Scanning." *Public Administration Review* 27.

Eyerman, R., and A. Jamison. 1991. *Social Movements: A Cognitive Approach*. Polity.

Finkel, A. 1996. "Who's Exaggerating?" *Discover*, May: 48–54.

Finucane, S. 1992. "Informant: Unocal Spills Span 13 Years." *Santa Barbara News Press*, July 24.

Finucane, S. 1997. "Unocal Preparing for El Niño at Tainted Beach Site." *Santa Barbara News Press*, November 11.

Finucane, S. 1998. "Unocal to Pay $43.8 Million." *Santa Barbara News Press*, July 22.

Fiorenza, T. 1994. "Corporate Crime: Crime Pays Just Fine—If You're Rich Enough." *San Luis Obispo New Times*, April 13.

Fischer, H. 1994. *Response to Disaster: Fact versus Fiction and Its Perpetuation.* Second edition. University Press of America.

Fischoff, B. 1990. "Understanding Long-Term Environmental Risks." *Journal of Risk and Uncertainty.* 3: 315–330.

Fishman, M. 1978. "Crime Waves as Ideology." *Social Problems* 25: 531–543.

Flacks, R. 1988. *Making History.* Columbia University Press.

Foran, J. 1997. "Discourses and Social Forces: The Role of Culture and Cultural Studies in Understanding Revolutions." In *Theorizing Revolutions*, ed. J. Foran. Routledge.

Foster, J. 1999. "Marx's Theory of Metabolic Rift: Classical Foundations for Environmental Sociology." *American Journal of Sociology* 105, no. 2: 366–405.

Fowlkes, M., and P. Miller. 1982. *Love Canal: The Social Construction of Disaster.* Federal Emergency Management Agency.

Freudenburg, W. 1988. "Perceived Risk, Real Risk: Social Science and the Art of Probabilistic Risk Assessment." *Science* 242, October: 44–49.

Freudenburg, W. 1992. "Nothing Recedes Like Success? Risk Analysis and the Organizational Amplification of Risk." *Risk* 3: 1–35.

Freudenburg, W. 1993. "Risk and Recreancy: Weber, the Division of Labor, and Rationality of Risk Perceptions." *Social Forces* 69: 1143–1168.

Freudenburg, W. 1994. Strange Chemistry: Environmental Risk Conflicts in a World of Science, Values, and Blind Spots. Keynote address, Environmental Risk Decision-Making: Values, Perceptions and Ethics, Annual Meeting, American Chemical Society, Washington.

Freudenburg, W. 2000. "Social Construction and Social Constructions: Toward Analyzing the Social Construction of 'the Naturalized' as well as the Natural." In *Environment and Global Modernity*, ed. G. Spaargaren et al. Sage.

Freudenburg, W., and R. Gramling. 1994. *Oil in Troubled Waters: Perceptions, Politics, and the Battle over Offshore Drilling.* State University of New York Press.

Freudenburg, W., and T. Jones. 1991. "Attitudes and Stress in the Presence of Technological Risk: A Test of the Supreme Court Hypothesus." *Social Forces* 69, no. 4: 1143–1168.

Freudenburg, W., and S. Pastor. 1992. "Public Responses to Technological Risks: Toward a Sociological Perspective." *Sociological Quarterly* 33, no. 3: 389–412.

Freudenburg, W., S. Frickel, and R. Gramling. 1995. "Beyond the Nature/Society Divide: Learning to Think about a Mountain." *Sociological Forum* 10, no. 3: 361–389.

Friesen, T. 1993. "Criminal Charges May Be Eliminated against Unocal." *Five Cities Times-Press-Recorder*, December 7.

Fritz, C. 1961. "Disaster." In *Contemporary Social Problems*, ed. R. Merton and R. Nesbit. Harcourt Brace & World.

Gamson, W., and A. Modigliani. 1989. "Media Discourse and Public Opinion on Nuclear Power: A Constructionist Approach." *American Journal of Sociology 95*: 1–37.

Gans, H. 1980. *Deciding What's News*. Vintage.

Garfinkel, H. 1956. "Conditions of Successful Degradation Ceremonies." *American Journal of Sociology* 61: 420–424.

Gibbs, L. 1982. *Love Canal: My Story*. State University of New York Press.

Gibbs, W. 1999. "Not Cleaning Up: Faster, Cheaper Ways to Restore Polluted Ground Are Largely Shunned." *Scientific American*, February: 39–41.

Giddens, A. 1990. *The Consequences of Modernity*. Stanford University Press.

Giddens, A. 1991. *Modernity and Self-Identity: Self and Society in the Late Modern Age*. Polity.

Giddens, A. 1994. "Living in a Post Traditional Society." In *Reflexive Modernization*, ed. U. Beck et al. Stanford University Press.

Gidney, C., B. Brooks, and E. Sheridan. 1917. *History of Santa Barbara, San Luis Obispo, and Ventura Counties, California*. Lewis.

Glantz, M., ed. 1999. *Creeping Environmental Problems and Sustainable Development in the Aral Sea Basin*. Cambridge University Press.

Glazer, M. 1987. "Whistleblowers." In *Corporate and Governmental Deviance*, ed. M. Ermann and R. Lundman. Third edition. Oxford University Press.

Goffman, E. 1959. *The presentation of Self in Everyday Life*. Doubleday.

Goldblatt, D. 1996. *Social Theory and the Environment*. Westview.

Gordon, A., and D. Suzuki. 1990. *It's a Matter of Survival*. Harvard University Press.

Gorz, A. 1980. *Ecology as Politics*. South End.

Gottlieb, R. 1993. *Forcing the Spring: The Transformation of the American Environmental Movement*. Island.

Gould, S. 1986. "Evolution and the Triumph of Homology, or Why History Matters." *American Scientist* 74: 60–69.

Gouldner, A. 1954. *Patterns of Industrial Bureaucracy*. Free Press.

Gramling, R. 1996. *Oil on the Edge: Offshore Development, Conflict, Gridlock*. State University of New York Press.

Granoetter, M. 1985. "Economic Action and Social Structure: The Problem of Embeddedness." *American Journal of Sociology* 91, no. 3: 481–510.

Greene, J. 1992a. "Oil Washes Ashore at Avila." *Telegram-Tribune* (San Luis Obispo), August 5.

Greene, J. 1992b. "Unocal: A Leaky Environmental Record." *Telegram-Tribune*, August 5.

Greene, J. 1993a. "Unocal Spills May Have Gone Unreported." *Telegram-Tribune*, July 1.

Greene, J. 1993b. "Unocal Workers Confirm Leaks." *Telegram-Tribune*, June 17.

Greene, J. 1994. "Unocal Accused of Breaking Honor Code." *Telegram-Tribune*, July 12.

Habermas, J. 1989. *The Structural Transformation of the Public Sphere*. Polity.

Hannan, M., and J. Freeman. 1984. "Structural Inertia and Organizational Change." *American Sociological Review* 49, April: 149–164.

Hanson, D. 1998. *Waste Land: Meditations on a Ravaged Landscape*. Aperture.

Haraway, D. 1991. *Simians, Cyborgs, and Women: The Reinvention of Nature*. Free Press.

Harr, J. 1995. *A Civil Action*. Vintage Books.

Hart, G. 1995. "How Unocal Covered Up a Record-Breaking California Oil Spill." In *The News That Didn't Make the News and Why*, ed. C. Jensen. Four Walls, Eight Windows.

Hasalm, G. 1972. *The Language of the Oilfields: Examination of an Industrial Argot*. Old Adobe.

Hawkins, K. 1983. "Bargain and Bluff: Compliance Strategy and Deterrence in the Enforcement of Regulation." *Law and Policy Quarterly* 5, no. 1: 35–73.

Heimer, C. 1985. *Reactive Risk and Relative Risk: Managing Moral Hazard in Insurance Contracts*. University of California Press.

Heimer, C. 1988. Social Structure, Psychology, and the Estimation of Risk. *Annual Review of Sociology* 14: 491–519.

Hewitt, K., ed. 1983. *Interpretations of Calamity: From the Viewpoint of Human Ecology*. Allen & Unwin.

Hilgartner, S. 1992. "Organizations, Uncertainties, and Risk." In *Organizations, Uncertainties, and Risk*, ed. J. Short Jr. and L. Clarke. Westview.

Hofrichter, R. 1993. *Toxic Struggles: The Theory and Practice of Environmental Justice*. New Society.

Hughes, T. 1985. "Edison and the Electric Light." In *The Social Shaping of Technology*, ed. D. MacKenzie and J. Wajcman. Open University Press.

Hummel, R. 1987. *The Bureaucratic Experience*. Third edition. St. Martin's.

Humphrey, C, and F. Buttel, 1982. *Environment, Energy, and Society*. Wadsworth.

Jackoway, R. 1990. "State Officials Differ on Oil Leak Damage." *Five Cities Times-Press-Recorder*, March 7.

Jacques, E. 1990. "In Praise of Hierarchy." *Harvard Business Review*. January-February: 127–133.

Jamison, A., R. Eyerman, and J. Cramer. 1990. *The Making of the New Environmental Consciousness: A Comparative Study of the Environmental Movements in Sweden, Denmark, and the Netherlands*. Edinburgh University Press.

Janis, I. 1982. *Groupthink: Psychological Studies of Policy Decisions and Fiascoes*. Houghton Mifflin.

Jepperson, R. 1991. "Institutions, Institutional Effects, and Institutionalism." In *The New Institutionalism in Organizational Analysis*, ed. W. Powell and P. DiMaggio. University of Chicago Press.

Jones, W. 1989. "Oil Spill Compensation and Liability Legislation: When Good Things Don't Happen to Good Bills." *Environmental Law Reporter* 10: 333–335.

Kagan, R., and J. Scholz. 1984. "'The Criminology of the Corporation' and Regulatory Enforcement Strategies." In Keith Hawkins and John M. Thomas, *Enforcing Regulations*. Kluwer-Nijhoff.

Kalberg, S. 1980. "Max Weber's Types of Rationality: Cornerstones for the Analysis of Rationalization Processes in History." *American Journal of Sociology* 85, no. 5: 1145–1179.

Kalberg, S. 1994. *Max Weber's Comparative Historical Sociology*. University of Chicago Press.

Kallman, R., and E. Wheeler. 1984. *Coastal Crude in a Sea of Conflict*. Blake.

Katz, J. 1977. "Cover-up and Collective Integrity: On the Natural Antagonisms of Authority Internal and External to Organizations." *Journal of Social Problems* 25, no. 1: 3–17.

King, A., and M. Lenox. 2000." Industry Self-regulation Without Sanctions: The Chemical Industry's Responsible Care Program. *Academy of Management Journal* 43, no. 4: 698–716.

Krieger, D. 1990. *San Luis Obispo County: Looking Backward into the Middle Kingdom*. Second edition. EZ Nature Books.

Kroll-Smith, S., and S. Couch. 1990. *The Real Disaster Is Above Ground: A Mine Fire and Social Conflict*. University Press of Kentucky.

Kroll-Smith, S., and S. Couch. 1991. "What Is a Disaster? An Ecological Symbolic Approach to Resolving the Definitional Debate." *International Journal of Mass Emergencies and Disasters* 9: 355–366.

Kroll-Smith, S., and S. Couch. 1993. "Symbols, Ecology, and Contamination: Case Studies in the Ecological-Symbolic Approach to Disaster." *Research in Social Problems and Public Policy* 5: 47–74.

Kuletz, V. 1998. *The Tainted Desert: Environmental and Social Ruin in the American West*. Routledge.

La Porte, T., and P. Consolini. 1991. "Working in Practice But not in Theory: Theoretical Challenges of High Reliability Organizations." *Journal of Public Administration Research and Theory*. January: 19–47.

Lash, S., B. Szerszynski, and B. Wynne. 1996. *Risk, Environment, and Modernity: Towards a New Ecology*. Sage.

Latour, B. 1987. *Science in Action*. Harvard University Press.

Latour, B. 1993. *We Have Never Been Modern*. Harvester Wheatsheaf.

Latour, B., and S. Woolgar. 1986. *Laboratory Life: The Construction of Scientific Facts*. Princeton University Press.

Law, J. 1987. "Technology and Heterogeneous Engineering: The Case of Portuguese Expansion." In *The Social Construction of Technological Systems*, ed. W. Bijker et al.. MIT Press.

Law, J. 1992. "Introduction: Monsters, Machines, and Sociotechnical Relations." In *A Sociology of Monsters*, ed. J. Law. Routledge.

Lee, G., V. Leon, L. MacDonald, et al. 1977. *An Uncommon Guide to San Luis Obispo County, California*. Padre Productions.

Lee, M., and D. Ermann. 1999. "'Pinto Madness' as Flawed Landmark Narrative: An Organizational and Network Analysis." *Social Problems* 46, no. 1: 30–47.

Levine, A. 1982. *Love Canal: Science, Politics, and People*. Lexington Books.

Levine, S., and P. White. 1961. "Exchange as a Conceptual Framework for the Study of Inter-Organizational Relationships." *Administrative Science Quarterly* 5: 583–601.

Lima, J. 1994. The Politics of Offshore Energy. Ph.D. dissertation, University of California, Santa Barbara.

Lindblom, C. "The Science of Muddling Through." *Public Administration Review* 19: 79–88.

Lindblom, C. 1979. "Still Muddling Through, Not Through Yet." *Public Administration Review* 39, no. 6: 517–526

Logan, J., and H. Molotch. 1987. *Urban Fortunes: The Political Economy of Place*. University of California Press.

Longino, H. 1990. *Science as Social Knowledge*. Princeton University Press.

Lowrance, W. 1976. *Of Acceptable Risk: Science and the Determination of Safety*. Kaufman.

MacNaghten, P., and J. Urry. 1995. "Towards a Sociology of Nature." *Sociology* 29, no.2: 203–220.

March, J. 1956. "Rational Choice and the Structure of the Environment." *Psychological Review* 63, no. 2.

March, J. 1978. "Bounded Rationality, Ambiguity, and the Engineering of Choice." *Bell Journal of Economics* 9, no. 2: 587–608.

March, J., and J. Olsen. 1979. *Ambiguity and Choice in Organizations*. Universitetsforlaget.

March, J., and H. Simon. 1958. *Organizations*. Wiley.

Marcuse, H. 1964. *One Dimensional Man: Studies in the Ideology of Advanced Industrial Society*. Beacon.

Marone, J., and E. Woodhouse. 1986. *Averting Catastrophe: Strategies for Regulating Risky Technologies*. University of California Press.

Martin, B. 1991. *Scientific Knowledge in Controversy*. State University of New York Press.

Marx, K. 1974. *Early Writings*. Vintage.

Mazur, A. 1973. "Disputes Between Experts." *Minerva* 11, no. 2: 243–262.

Mazur, A. 1975. "Opposition to innovation." *Minerva* 13, no. 1: 58–81.

Mazur, A. 1981. *The Dynamics of Technical Controversy.* Communications Press.

Mazur, A. 1991. "Putting Radon and Love Canal on the Public Agenda." In *Communities at Risk*, ed. S. Couch and S. Kroll-Smith. Lang.

Mazur, A. 1998. *A Hazardous Inquiry: The Rashomon Effect at Love Canal.* Harvard University Press.

McGinnis, M. 1991. "San Luis Obispo's Measure A: Check or Checkmate?" In *The California Coastal Zone Experience*, ed. G. Domurat and T. Wakeman. American Society of Civil Engineers.

McGinnis, M., and J. Proctor. 1999. "Tragic Choice in Bio-diversity Conservation Policy." *Society and Natural Resources* 34: 563–588.

McKee, J., and H. Wolf, eds. 1963. Water Quality Criteria. Publication 3-A, California Water Quality Control Board.

McLellan, David. 1977. *Karl Marx: Selected Writings.* Oxford University Press.

McMahon, J. 1994a. "Unocal's Other Spill: While Guadalupe Dunes Steals the Headlines, Another Huge Underground Spill Goes Unnoticed and Unattended to in San Luis Obispo." *San Luis Obispo New Times*, April 13.

McMahon, J. 1994b. "Unocal Keeps Popping Up on the Underground Spills List." *San Luis Obispo New Times*, April 13.

Melucci, A. 1989. *Nomads of the Present: Social Movements and Individual Needs in Contemporary Society.* Radius.

Merchant, C. 1980. *The Death of Nature.* Harper and Row.

Merton, R. 1936. "The Unanticipated Consequences of Purposive Social Action." *American Sociological Review* 1, no. 6: 894–904.

Merton, R. 1940. "Bureaucratic Structure and Personality." *Social Forces* 18: 560–568.

Meyer, J., and B. Rowan. 1977. "Institutionalized Organizations: Formal Structure as Myth and Ceremony." *American Journal of Sociology* 83, no. 2: 340–363.

Milgram, S. 1974. *Obedience to Authority.* Harper & Row.

Mitchell, J. 1990. "Human Dimensions of Environmental Hazards: Complexity, Diparity, and the Search for Guidance." In *Nothing to Fear*, ed. A. Kirby. University of Arizona Press.

Molotch, H. 1970. "Oil in Santa Barbara and Power in America." *Sociological Inquiry* 40 (Winter): 131–144.

Molotch, H. 1979. "Media and Movements." In *The Dynamics of Social Movements*, ed. M. Zald and J. McCarthy. Winthrop.

Molotch, Harvey. 1990. "Sociology, Economics, and the Economy." In *Sociology in America*, ed. H. Gans. Sage.

Molotch, H., and W. Freudenburg. 1996. Santa Barbara County: Two Paths. Final Report. OCS Study MMS96-0037. Department of Interior, Minerals Management Service, Pacific OCS Region. Camarillo, Calif.: Ocean Coastal Policy Center, Marine Science Institute, University of California, Santa Barbara.

Molotch, H., and M. Lester. 1975. "Accidental News: The Great Oil Spill as Local Occurrence and National Event." *American Journal of Sociology* 81: 235–260.

Molotch, H., W. Freudenburg, and K. Paulsen. 2000. History Repeats Itself, But How? City Character, Urban Tradition, and the Accomplishment of Place. *American Sociological Review* 65, no. 6: 791–823.

Morone, J., and E. Woodhouse. 1986. *Averting Catastrophe: Strategies for Regulating Risky Technologies*. University of California Press.

Mumford, L. 1934. *Technics and Civilization*. Harcourt Brace.

Mumford, L. 1964. *Pentagon of Power: The Myth of the Machine*. Harcourt Brace Jovanovich.

Nash, R. 1989. *The Rights of Nature: A History of Environmental Ethics*. University of Wisconsin Press.

National Oceanic and Atmospheric Administration. 1992. "Oil Spill Case Histories 1967–1991: Summaries of Significant US and International Spills." Report HMRAD 92-11. Hazardous Materials, Response, and Assessment Division, Seattle.

Nelson, R., and S. Winter. 1982. *An Evolutionary Theory of Economic Change*. Harvard University Press.

Nevarez, L., H. Molotch, and W. Freudenburg. 1996. A Major Switching. Final Report. OCS Study MMS96-0037. Department of Interior, Minerals Management Service, Pacific OCS Region. Camarillo, Calif.: Ocean Coastal Policy Center, Marine Science Institute, University of California, Santa Barbara.

Olsen, D. 1986. "South County Oil Lure." *South County Tribune*, November 6.

Olsen, D. 1991. "Silliman Fueled Great Debate on County Oil." *South County Tribune*, July 20.

Paddock, R. 1994a. "Painstaking Efforts Expose State's Largest Oil Spill." *Los Angeles Times*, March 21.

Paddock, R. 1994b. "Setbacks Slow Cleanup of Huge Oil Spill." *Los Angeles Times*, October 11.

Paulsen, K., H. Molotch, and W. Freudenburg. 1996. Oil, Fruit, Commune and Commute. Final Report. OCS Study MMS96-0037. Department of Interior, Minerals Management Service, Pacific OCS Region. Camarillo, Calif.: Ocean Coastal Policy Center, Marine Science Institute, University of California, Santa Barbara.

Pentland, B., and H. Rueter. 1994. "Organizational Routines as Grammars of Action." *Administration Science Quarterly* 39: 484–510.

Perrow, C. 1984. *Normal Accidents: Living With High Risk Technologies*. Basic Books.

Perrow, C. 1986. *Complex Organizations: A Critical Essay*. McGraw-Hill.

Perrow, C. 1991. "A Society of Organizations." *Theory and Society* 20: 725–762.

Perrow, C. 1997. "Organizing for Environmental Destruction." *Organization and the Environment* 10, no. 1: 66–72.

Perrucci, R., R. Anderson, D. Schendel, and L. Trachman. 1980. "Whistle-Blowing: Professionals' Resistance to Organizational Authority." *Social Problems* 28, no. 2: 149–164.

Pfaff, D. 1994. "Fear and the 'Silent Spill.'" *San Francisco Daily Journal*, June 23.

Powell, W., and P. DiMaggio, eds. 1991. *The New Institutionalism in Organizational Analysis*. University of Chicago Press.

Pratt, J. 1978. "Growth or a Clean Environment? Responses to Petroleum-related Pollution in the Gulf Coast Refining Region." *Business History Review* 52, no. 1: 1–29.

Pratt, J. 1980. "Letting the Grandchildren Do It: Environmental Planning During the Ascent of Oil as a Major Energy Source." *Public Historian* 2, no. 4.: 28–61.

Pressman, J., and A. Wildavsky. 1984. *Implementation: How Great Expectations in Washington Are Dashed in Oakland*. University of California Press.

Quam-Wickham, N. 1994. Petroleocrats and Proletarians: Work, Class, and Politics in the California Oil Industry, 1917–1925. Dissertation, University of California, Berkeley.

Quarantelli, E., ed. 1978. *Disaster Theory and Research*. Sage.

Quarantelli, E., and R. Dynes. 1972. Images of Disaster Behavior: Myths and Consequences. Disaster Research Center, University of Delaware.

Quarantelli, E., and R. Dynes. 1977. "Responding to Social Crises and Disaster." *Annual Review of Sociology* 3: 305–312.

Randall, J. 1976. *The Making of the Modern Mind*. Columbia University Press.

Rayner, S. 1987. Risk and Relativism in Science and Policy." In *The Social and Cultural Construction of Risk*, ed. B. Johnson and V. Covello. Reidel.

Rayner, S., and R. Cantor. 1987. "How Fair Is Safe Enough? The Cultural Approach to Societal Technology Choice." *Society for Risk Analysis* 7: 3–9.

Redclift, M., and G. Woodgate, eds. 2000. *The International Handbook of Environmental Sociology*. Elgar.

Reed, J.-P., and J. Foran. 1999. Political Cultures of Opposition: Exploring Idoims, Ideologies, and Revolutionary Agency. Manuscript, Department of Sociology, University of California, Santa Barbara.

Reinarman, C. 1987. *American States of Mind: Political Beliefs and Behavior Among Private and Public Workers*. Yale University Press.

Rice, A. 1994. "Endless Bummer." *Santa Barbara Independent*, March 17.

Ritea, S. 1994. "Silent Spill." *San Luis Obispo New Times*, February 23.

Roberts, K. 1982. "Characteristics of One Type of High Reliability Organization." *Organizational Science* 1, no. 2: 160.

Romo, J. 2000. Rigging Genders: The Structure of Masculinity in Offshore Oil Production. Master's thesis, Department of Sociology, University of California, Santa Barbara.

Sagan, S. 1993. *The Limits to Safety: Organizations, Accidents, and Nuclear Weapons*. Princeton University Press.

San Luis Obispo County Planning Department. 1995. Guadalupe Oilfield Remediation Project Newsletter, July 1.

San Luis Obispo District Court. 1993. Prosecutions Witness, Deposition Taken Under Oath. Case M-202983, December 20.

Saxenian, A. 1994. *Regional Advantage: Culture and Competition in Silicon Valley and Route 128*. Harvard University Press.

Schnaiberg, A. 1980. *The Environment: From Surplus to Scarcity*. Oxford University Press.

Schnaiberg, A., and K. Gould. 1994. *Environment and Society: The Enduring Conflict*. St. Martin's.

Schulman, P. 1989. "The Logic of Organizational Irrationality." *Administration and Society* 21, no. 1: 31–53.

Scott, R.1981. *Organizations: Rational, Natural, and Open Systems*. Prentice-Hall.

Sherman, L. 1987. "Deviant Organizations." In *Corporate and Governmental Deviance*, ed. M. Ermann and R. Lundman. Third edition. Oxford University Press.

Short, J., and L. Clarke, eds. 1992. *Organizations, Uncertainties, and Risk*. Westview.

Shove, E., L. Lutzenhiser, B. Hackett, S. Guy, and H. White. 1998. "Energy and Social Systems." In *Human Choice and Climate Change*, ed. S. Rayner et al. Battelle.

Shrader-Frechette, K. 1991. *Risk and Rationality: Philosophical Foundations for Populist Reforms*. University of California Press.

Shrader-Frechette, K. 1994. "Science, Environmental Risk Assessment, and the Frame Problem." *BioScience* 44, no. 8: 548–551.

Shrader-Frechette, K. 1995. "Evaluating the Expertise of the Experts." *Risk: Health, Safety and Environment* 115 (Spring): 115–126.

Shrader-Frechette, K. 1997. "Elite Folk Science and Environmentalism." *Organizations and the Environment* 10, no. 1: 23–35.

Shrader-Frechette, K., and E. McCoy. 1993. *Method in Ecology*. Cambridge University Press.

Simon, H. 1947. *Administrative Behavior*. Second edition. Macmillan.

Simon, H. 1955. "A Behavioral Model of Rational Choice." *Quarterly Journal of Economics* 69, February.

Simon, H. 1956. "Rational Choice and the Structure of the Environment." *Psychological Review* 63, no. 2: 129–139.

Skillern, F. 1981. *Environmental Protection: The Legal Framework*. McGraw-Hill.

Slater, D. 1994. "Dress Rehearsal for Disaster." *Sierra Magazine*, May-June: 53.

Slovic, P. 1993. "Perceived Risk, Trust, and Democracy." *Risk Analysis* 13, no. 6: 675–682.

Slovic, P., and B. Fischhoff. 1983. "How Safe Is Safe Enough? Determinants of Perceived and Acceptable Risk." In *To Hot to Handle?* ed. L. Gould et al. Yale University Press.

Slovic, P., B. Fischhoff, and S. Lichtenwtein. 1979. "Rating Risks: The Structure of Expert and Lay Perceptions." *Environment* 21, no. 3: 14–39.

Sneed, D. 1997. "Guadalupe Dunes' Oil Pollution: How Bad?" *Telegram-Tribune*, February 20.

Sneed, D. 1998. "Coastal Official Tells Unocal to Clean Up." *Telegram-Tribune*, January 13.

Sneed, D. 1999. "Guadalupe Cleanup Gets Go-Ahead." *Telegram-Tribune*, November 4.

Snook, S. 2000. *Friendly Fire: The Accidental Shootdown of US Black Hawks over Northern Iraq*. Princeton University Press.

Solen, R. 1998. *An Ocean of Oil: A Century of Political Struggle over Petroleum off the California Coast*. Denali.

Stallings, R. 1990. "Media Discourse and the Social Construction of Risk." *Social Problems* 37, no. 1: 80–95.

Stanley, M. 1968. "Nature, Culture, and Scarcity: Foreword to a Theoretical Synthesis." *American Sociological Review* 33, no. 6: 855–870.

Starr, C. 1969. "Social Benefit versus Technological Risk." *Science* 165: 1232.

Staw, B. 1981. "The Escalation of Commitment to a Course of Action." *Academy of Management Review* 6, no. 4: 577–587.

Staw, B., and J. Ross. 1989. "Understanding Behavior in Escalating Situation." *Science* 246, no. 4927: 216.

Stinchcombe, A. 1965. "Social Structure and Organizations." In *Handbook of Organizations*, ed. J. March. Rand McNally.

Stormont, D. 1956. "Diluents Permit Faster Pumping Speeds . . . and Give Other Production Benefits." *Oil and Gas Journal*, August 13: 127–128.

Storper, M., and R. Salais. 1997. *Worlds of Production: The Action Frameworks of the Economy*. Harvard University Press.

Stover, M. 1989a. "Oil Slump: Decline in County Drilling Means Losses of Jobs, Taxes." *Telegram-Tribune*, January 26.

Stover, M. 1989b. "Oil Spill: What If It Happened Here?" *Telegram-Tribune*, April 26.

Superior and Municipal Courts of the State of California County of San Luis Obispo. 1998. People of the State of California ex. rel. California Department of Fish and Game et al., Plaintiffs, v. Union Oil Company of California dba Unocal, and California Corporation et al., Defendants. Settlement Agreement and Judgement cv75194, July 22.

Szazs, A. 1994. *Ecopopulism: Toxic Waste and the Movement for Environmental Justice*. University of Minnesota Press.

Szerszynski, B. 1996. "On Knowing What to Do: Environmentalism and the Modern Problematic." In *Risk, Environment and Modernity*, ed. S. Lash et al. Sage.

Tenner, E. 1998. *Why Things Bite Back: Technology and the Revenge of the Unintended Consequences*. Vintage Books.

Tester, K. 1991. *Animals and Society: The Humanity of Animal Rights*. Routledge.

Thompson, J. 1967. *Organization in Action*. McGraw-Hill.

Tierney, K. 1999. "Toward a Critical Sociology of Risk." *Sociological Forum* 14, no. 2: 215–242.

Turner, B. 1971. *Exploring the Industrial Subculture*. Macmillan.

Turner, B. 1978. *Man-Made Disasters*. Wykeham.

Turner, J., and J. Rylander. 1997. "Land Use: The Forgotten Agenda." In *Thinking Ecologically*, ed. M. Chertow and D. Esty. Yale University Press.

Tversky, A., and D. Kahneman. 1974. "Judgement under Uncertainty: Heuristics and Biases." *Science* 185: 1124–1131.

Unocal Corporation. 1993. "To Members of the Central Coast Community." *Telegram-Tribune*, July 29.

Unocal Corporation. 1994a. Environmental Report to Stockholders.

Unocal Corporation. 1994b. Guadalupe Project: A Progress Report. Corporate Communications Department, Seventy Six Special Edition, Los Angeles.

Unocal Corporation. 1996a. Yesterday Tomorrow Today: Unocal's Commitment to the Community. Unocal Corporation.

Unocal Corporation. 1996b. "Yesterday Tomorrow Today: Unocal's Commitment to the Community." *San Luis Obispo New Times*, October 31.

Unocal Corporation. 1997. "A New World, a New Unocal: 1996 Annual Report." El Segundo: Union Oil Company of California.

US Coast Guard. 1995a. Appendix E of the National Oil and Hazardous Substances Pollution Contingency. Office of Marine Safety, Security, and Environmental Protection (reprint title: "Appendix to the National Contingency Plan: Oil Spill Response"). (January 20).

US Coast Guard. 1995b. Incident Specific Preparedness Review Final Report of the Response to the Oil Spill Resulting From Leakage of Diluent Oil from Shorelines of Guadalupe Beach, CA. Mobile: US Department of Transportation. (August 2).

US District Court, Northern California. 1994. Attorney for Plaintiff Brief, Environmental Center of San Luis Obispo County, CA, Defendants, Union Oil Company of California, United States Coast Guard, California Coastal Commission, and County of San Luis Obispo, CA. Court Transcripts, Plaintiff's Allegation Taken Under Oath. (September 2): 3.

US Nuclear Regulatory Commission. 1975. Reactor Safety Study: An Assessment of Accident Risks in US Commercial Nuclear Power Plants.

Vaughan, D. 1996. *The Challenger Launch Decision: Risky Technology, Culture, and Deviance at NASA*. University of Chicago Press.

Vaughan, D. 1998. "Rational Choice, Situated Action, and Social Control of Organizations." *Law and Society Review* 32, no. 1: 23–61.

Vaughan, D. 1999. "The Dark Side of Organizations: Mistake, Misconduct, and Disaster." *Annual Review of Sociology* 25: 271–305.

Vyner, H. 1988. *Invisible Trauma: The Psychosocial Effects of Invisible Environmental Contaminants*. Lexington Books.

Walker, R. 1998. California's Debt to Nature: Natural Resources and the Golden Road to Capitalist Growth, 1848–1940. Manuscript, Department of Geography, University of California, Berkeley.

Walsh, E. 1981. "Resource Mobilization and Citizen Protest in Communities around Three Mile Island." *Social Problems* 26: 1–21.

Walsh, E. 1988. "Challenging Official Risk Assessment via Protest Mobilization: The TMI Case." In *The Social Construction of Risk*, ed. B. Johnson and V. Covello. Reidel.

Walton, J. 1992a. "Making a Theoretical Case." In *What Is a Case?* ed. C. Ragin and H. Becker. Cambridge University Press.

Walton, J. 1992b. *Western Times and Water Wars: State, Culture, and Rebellion in California*. University of California Press.

Warren, R., S. Rose, and A. Bergunder. 1974. *The Structure of Urban Reform: Community Decision Organizations in Stability and Change*. Lexington Books.

Weale, A. 1992. *The New Politics of Pollution*. Manchester University Press.

Weber, M. 1952. *The Protestant Ethic and the Spirit of Capitalism*. Scribner.

Weber, M. 1968. *Economy and Society*. Bedminister.

Weick, K. 1976. "Educational Organizations as Loosely Coupled Systems." *Administrative Science Quarterly* 21, March: 1–19.

Weick, K. 1979. *The Social Psychology of Organizing*. Second edition. Addison-Wesley.

Weick, K. 1987. "Culture as a Source of High Reliability." *California Management Review* 29, no. 2: 116.

Weick, K. 1993. "The Collapse of Sensemaking in Organizations: The Mann Gulch Disaster." *Administrative Science Quarterly* 38: 628–652.

Weick, K. 1995. *Sensemaking in Organizations*. Sage.

Weick, K., and K. Roberts. 1993. "Collective Mind in organizations: Heedful Interrelating on Flight Decks." *Administrative Science Quarterly* 38: 357–381.

Welty, E., and F. Taylor. 1958. *The Black Bonanza*. Second edition. McGraw-Hill.

Weston, A. 1981. *Whistleblowing: Loyalty and Dissent in the Corporation*. McGraw-Hill.

White, G. 1962. *Formative Years in the Far West: A History of Standard Oil Company of California and Predecessors through 1919*. Appleton-Century-Crofts.

White, K. 1993. "I Could No Longer Keep Silent." *Santa Maria Times*, July 22.

Wilcox, D. 1994a. "Regulators Blew It." *Telegram-Tribune*, January 5.

Wilcox, D. 1994b. "Unocal Case Setback: Prosecutors Missed Deadline to File Charges by Two Days, Judges Rules." *Telegram-Tribune*, January 5.

Wilcox, D. 1994c. "Unocal Pays $1.5 Million: A Timeline on the Guadalupe Disaster." *Telegram-Tribune*, March 16.

Wilcox, D. 1994d. "Unocal Cleanup in San Luis Obispo in Doubt." *Telegram-Tribune*, November 6.

Wildavsky, A. 1979. "No Risk Is the Highest Risk of All." *American Scientist* 67: 32–37.

Wildavsky, A. 1988. *Searching for Safety*. Transaction.

Wilkins, Le. 1987. "Risk Analysis and the Construction of News." *Journal of Communication* 37, no. 3: 80–92.

Williams, J. 1997. *Energy and Making of the Modern California*. University of Akron Press.

Wolf, S. 1988. *Pollution Law Handbook: A Guide to Federal Environmental Law*. Quorum Books.

Woodgate, G. 2000. "Introduction." In *The International Handbook of Environmental Sociology.*, ed. M. Redclift and G. Woodgate. Elgar.

Wynne, B. 1987. *Risk Management and Hazardous Wastes: Implementation and the Dialectics of Credibility*. Springer.

Wynne, B. 1992. "Misunderstood Misunderstanding: Social Identities and the Public Uptake of Science." *Public Understanding of Science* 1, no. 3: 281–304.

Wynne, B. 1996. "May the Sheep Safely Graze? A Reflexive View of the Expert-Lay Knowledge Divide." In *Risk, Environment and Modernity*, ed. S. Lash et al. Sage.

Yeager, P. 1991. *The Limits of Law: The Public Regulation of Private Pollution*. Cambridge University Press.

Yergin, D. 1991. *The Prize: The Epic Quest for Oil, Money and Power*. Simon & Schuster.

Index

Printed in the United States
By Bookmasters